The Awakening of Numbers

by Mitchell Freeman

The contents of this work, including, but not limited to, the accuracy of events, people, and places depicted; opinions expressed; permission to use previously published materials included; and any advice given or actions advocated are solely the responsibility of the author, who assumes all liability for said work and indemnifies the publisher against any claims stemming from publication of the work.

All Rights Reserved
Copyright © 2021 by Mitchell Freeman

No part of this book may be reproduced or transmitted, downloaded, distributed, reverse engineered, or stored in or introduced into any information storage and retrieval system, in any form or by any means, including photocopying and recording, whether electronic or mechanical, now known or hereinafter invented without permission in writing from the publisher.

Dorrance Publishing Co
585 Alpha Drive
Suite 103
Pittsburgh, PA 15238
Visit our website at *www.dorrancebookstore.com*

ISBN: 978-1-6491-3124-9
eISBN: 978-1-6491-3631-2

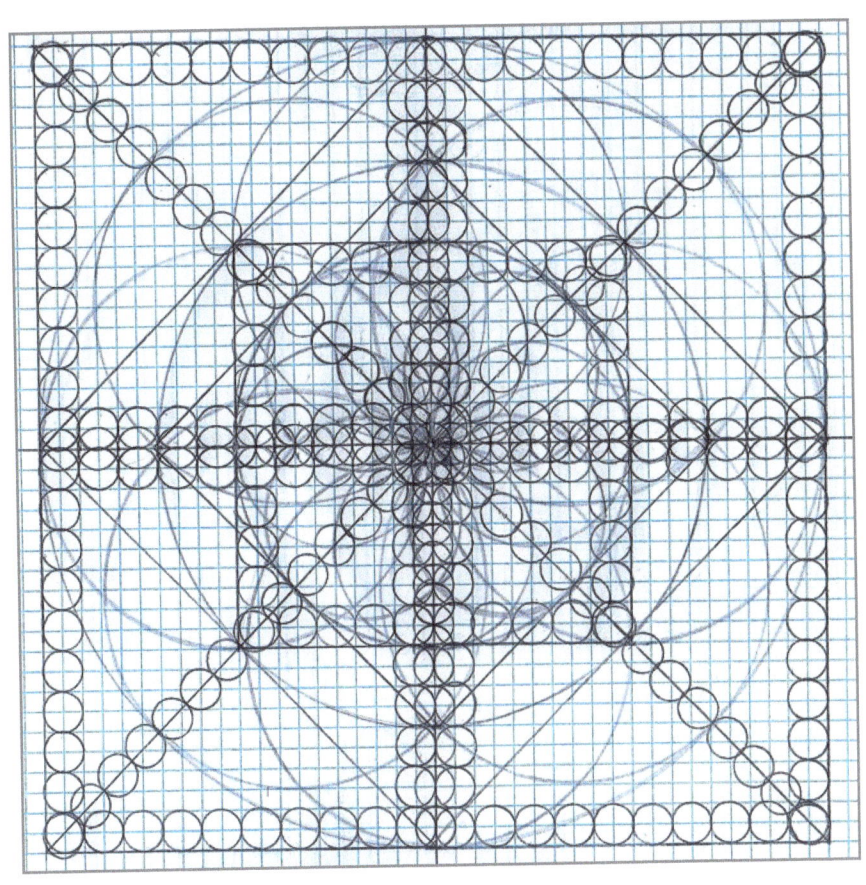

The Awakening of Numbers

Contents

Preface vii

Chapter 1: Originations. 1
Chapter 2: Arithmetic 15
Chapter 3: Numbers 31
Chapter 4: Circular Counting 41
Chapter 5: The Making of the Square Root of Two. 49
Chapter 6: Trigonometric Identities 59
Chapter 7: Number Uses. 75
Chapter 8: Pi (π), Protractor, Abacus 97
Chapter 9: Illustrations 119

Preface to Numbers

There are as many Number Systems as there are Numbers. This book is about the Numbers that make the Decimal System.

Have you ever sat in a Mathematics class and thought this makes no sense? So, you memorize what you are being told and repeat what you remember, when you can. This is the cause of estimations, guessing, rounding of numbers and randomness. There is none of that in Numbers, Arithmetic or Mathematics or this book. There is only Multiplication with Division after Multiplication, Addition with Subtraction after Addition, Squaring and Un-squaring (Square Root). This book will give you the ability to make sense of the Numbers and how they are manipulated, and at times misused.

Constructs (Proofs) will be used to demonstrate what makes sense and what does not. These constructs will also show how Numbers were developed and passed down through the Generations.

All things use numbers. Probably not as how Humans think of them or how Humans use them. It is evident that this is so if one just looks around at what is in this Universe. Some creatures and things make something, much as Humans do. For example: Birds make Nests, Bees make Honeycombs, Snails make shells, Atoms make radiation, and so on. The main difference is that Humans write it down. They do so to understand things and to make things and then pass this knowledge to other Humans.

Numbers are Circles, which are later found to be Spheres. The Circles may be put into a Square, later found to be a Cube, a Sphere may be put into a Cube. The Square may be Circled, and the Cube may be put in a Sphere. This book is about Decimal Numbers, Arithmetic and Basic Mathematics (Algebra, Trigonometry, and Geometry). The Mathematics associated with Spheres and Cubes will only be mentioned in this book and will not be analyzed in depth.

Arithmetic and Mathematics are used with Numbers to make the unknown, known.

Chapter 1
The Originations

First-Generation Numbers:

The Sun grew from the Earth, the seed from which it grew gave the Sun a root. Humans used rods and rope to measure it as it sat upon the Earth's Horizon. The rods may have had lengths that equated to the height and width of the rising Sun, with the center designated by rods or rope that extend from the top of one rod to the bottom of the opposite rod. A square was formed. This is a Zero-based One, all Zero-based numbers are formed from this One and all are identical (Figure 1).

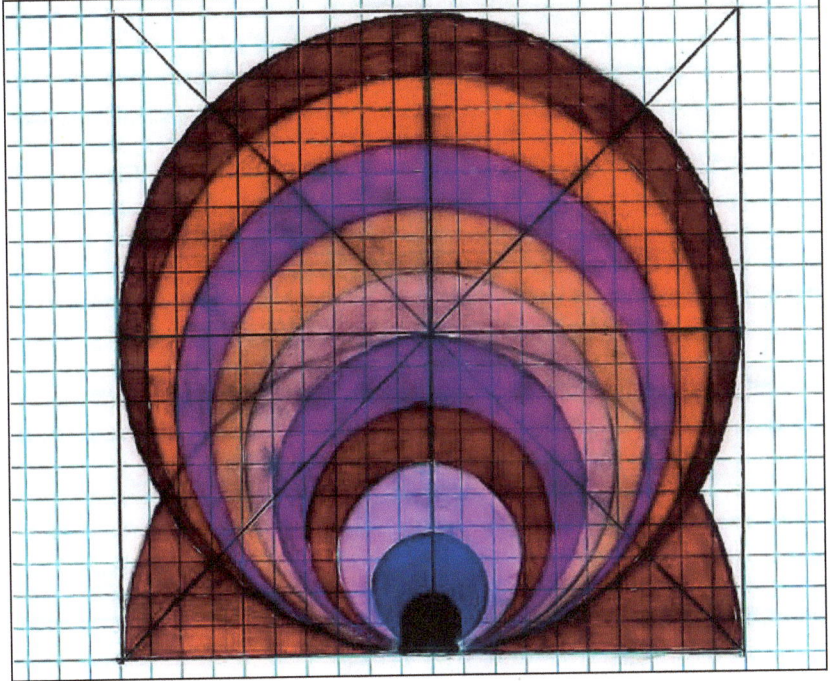

Figure 1

First-Generation numbers are a product of tracking the movement of the Sun (One) upon the Earth. Patterns were drawn on the earth. The motion of the Sun (One) on the earth, from Sunrise to Sunset was recorded (Figure 1a).

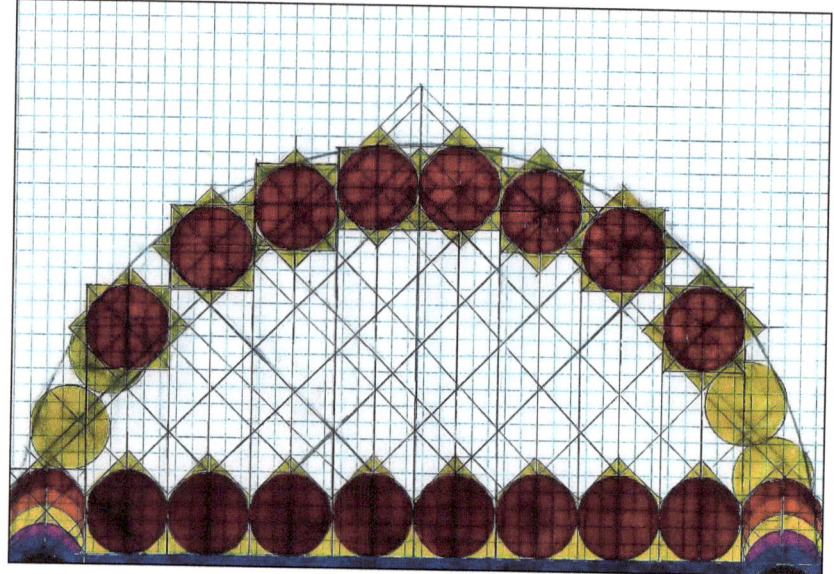

Figure 1a

Counting was performed with these Numbers, Human fingers and marks were probably used; the marks may have looked like this. I, II, III, IIII, IIIII, IIIII, IIII, III, II, I. The One grows from the ground from Sunrise until Sunset. This count is, 5 up from Sunrise to the peak of the Sun, and 5 down from the peak to Sunset.

Counting:

All Numbers Are Symbols. What the Symbol looks like depends on the time in history and the culture of the Humans who are using the symbols. The Zero is a line that represents a limit. It does not indicate that Nothing exists, it is a symbol that means there are no Numbers. Lines and contact points are counted. Counted lines may be Horizontal, Vertical or Diagonal. The intersection of two or more lines is designated as a point (.), the joining of two lines is designated as a point, the contact of a circle or arc with a line is designated as a point.

Second-Generation Numbers:

Second-Generation numbers are the result of using lines to describe the movement of the Sun above the Earth and the corresponding movement on the Earth. This is the beginning of Trigonometry. Lines are counted, and circles are assigned Number symbols. At this point in time the Sun is a disc (circle) and the Earth is still flat and there is only one Number and it is a Circle in a Square. (Figure 2). The center of the Sun (One) is tracked along the diagonal that is formed by the line extending from Zero to the top of the first vertical line.

This construct shows that counting is still performed. It is still 5 up and 5 down. The counting is performed by sliding and stacking of the squares. Zero is the root for the number One. Step and Equilateral Triangular Pyramid shapes are evident (Figure 2).

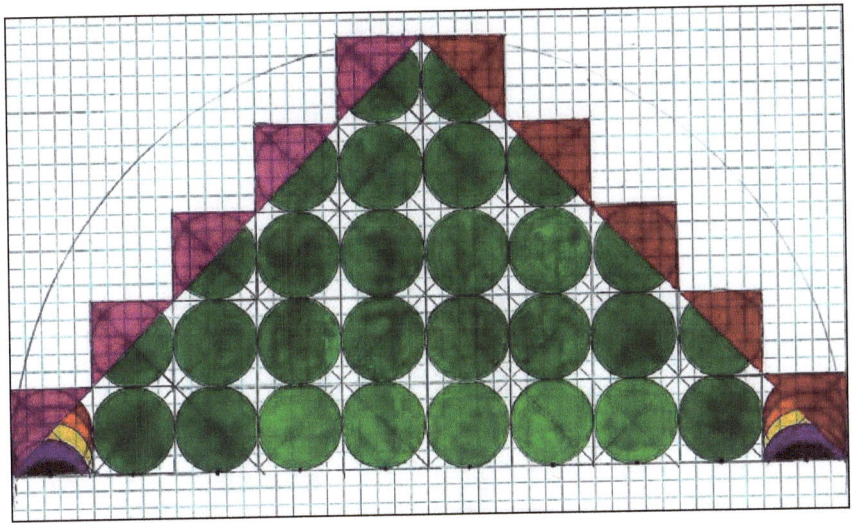

Figure 2

For ease in understanding, Number symbols, that are currently in use today, are assigned to each line. Zero means that no number is present on the line, it is Zero (0) (Figure 3).

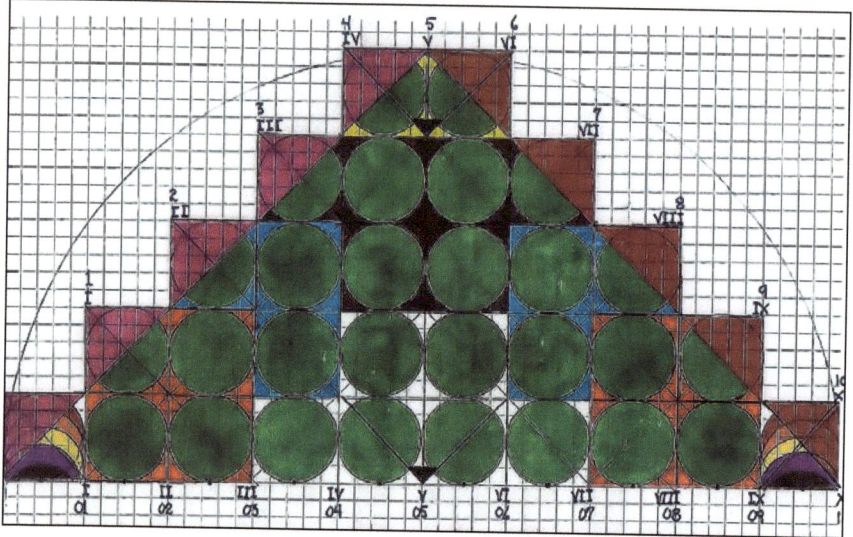

Figure 3

The first Number symbols we use are those referred to as Roman Numerals. The One ascends the Step Pyramid, the numbers increase by one until the top plateau is reached. In the middle of plateau is a line that intersects a V shape. As we move across the plateau the V increases by one. The count descends the step Pyramid until the One no longer exists, it is X'ed out. The same symbols are applied to the Horizontal counting lines.

We now assign the Number symbols, for the Decimal system, currently in use. Going up and down the Step Pyramid, Number symbols are assigned as shown. The horizontal line, is numbered as follows: zero to one = 01, zero to 2 = 02 and on to 10. Counting is now Vertical and Horizontal. The vertical lines are counted from left to right, indicating that the number has been used. The Number symbol 10 means that all the previous single-digit numbers (squares) have been used once and the count starts at zero again, with the appropriate single-digit number added, i.e.: 1.0, 10+, 20+, 30+ to Infinity.

The Second-Generation Number Construct was Horizontally aligned (Figure 4). Counting is performed and is like Abaci in use today.

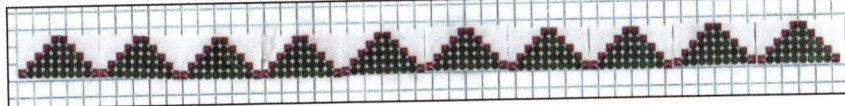

Figure 4

Horizontally aligned Second-Generation number constructs were then rotated to form a Counting table (Figure 5). The beginning and end of 4 days are now connected.

Figure 5

The construct shown in Figure 6 serves as the basis of all succeeding Counting table constructs. A Number is defined by the Counting Table a Number does not define the Counting Table. Several Arithmetical and Mathematical functions are derived from the Second-Generation Number Counting table.

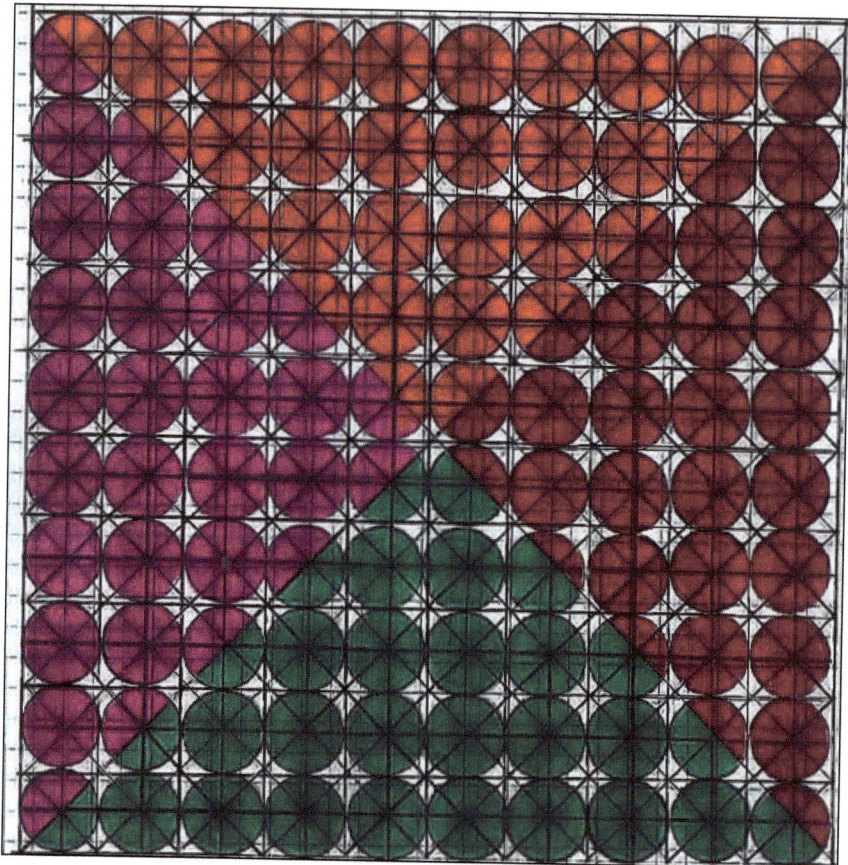

Figure 6

Squaring and Area are introduced at this time.

The counting table is comprised entirely of One units. There are no values assigned to the One unit until a number is transferred from another Table, there is only counting. For example: If one is assigned to the Table the One units will be .1 each. If ten is assigned to the

Table, each one unit will equal 1. If 2 is assigned to the table each One unit has a value of .2. This sequence may continue to Infinity.

The Ones along the sides are the source of all numbers on the Counting Table, meaning One is the Root of all the other numbers in this Construct and its combinations. The Root of the One is Zero. The first square after 1 is 2, horizontally and vertically. A 2 by 2 square is 2 columns of 2 or rows of 2 across, the count is 4 squares total. The squaring continues to 10 by 10 which yields a count of 100. (Figure 7)

Rectangles may also be described. For example: a 2 by 3 rectangle is formed by 2 vertical squares and 3 horizontal squares. The count of the squares is 6. Or a 3 by 2 rectangle is formed 3 vertical squares and 2 horizontal squares. The count is 6.

This analysis results in the equation. Area = Length by Width, and results in the first Algebraic Equation of: $A = L \times W$.

This counting results in the Arithmetical function known as Multiplication and is symbolized by the X that is within the square. Multiplication is always a move toward Infinity.

After Multiplication counting, Division counting can occur and is symbolized by the /, the diagonal line that passes through the ones within a square. Using Figure 7, we find that one side multiplied by a second side or 10 by (X, times) 10 has a count of 100 within a square. The area equation begins at $1 \times 1 = 1$.

Division is now possible, and the equation is: $A = C/B$, $A = 100/10 = 10$. The Division can continue to 1 and results in $1 / 1 = 1$. (Figure 7) Division is always a move toward 1

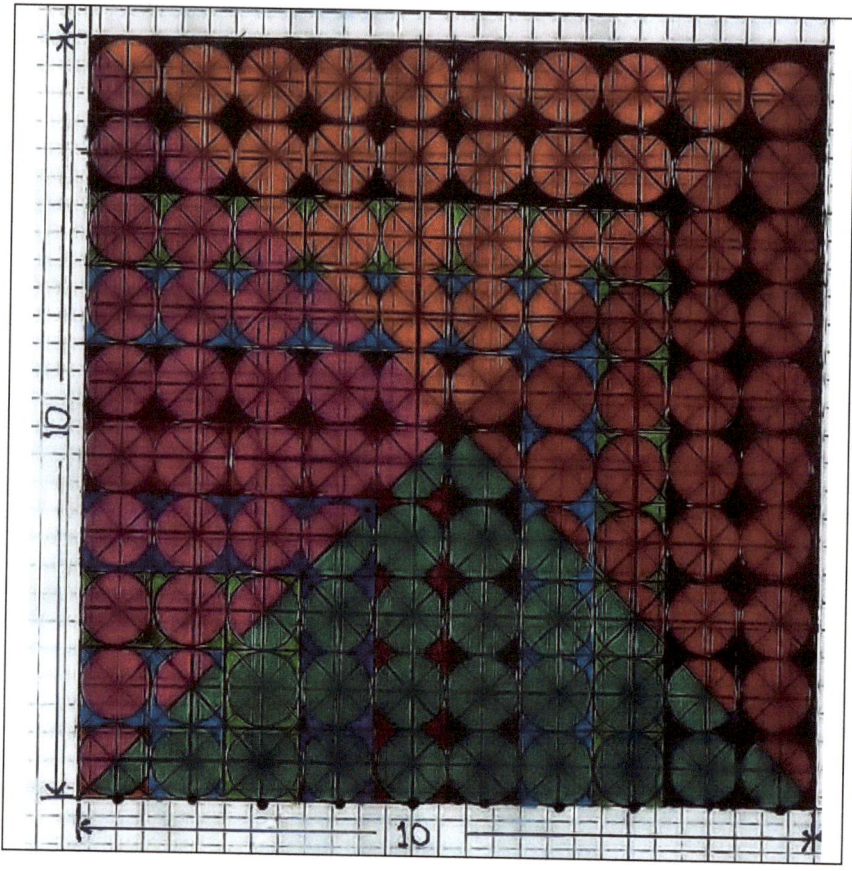

Figure 7

Squares and rectangles can be generated from any side of the square. There are no Negative numbers or a Fractured One (Fractions), there is only One (Whole Number). (Figure 7a)

Shown is 1 x 1, 1 X 2, 2 X 2, 2 X 4, 3 X 3, 3 X 5, and 4 X 4. With the exception of 1 X 1 all other numbers are considered a set, or a group of sets that comprise the number. On this counting table there is only the number One, so all sets are comprised of a one. When we multiply it is still A by (X, times) B = C. 1 X 1 = 1, 1 X 2 = 2, 2 X 2 = 4, 2 X 4 = 8, 3 X 3 = 9, 3 X 5 = 15 and 4 X 4 = 16. Division is now counting the number of times a specific number set is contained within the larger number, it is symbolized by a horizontal line , with the entire larger number on top of the line and the number set below the line. The symbol / is also used to denote division. There are 16 ones in a 4 X 4 square. Divide 16 by 2. There are 8 number 2 sets. Divide 16 by 3. There are 5 sets of 3 (Horizontal and Vertical) with a remainder of 1. This is division, and it should be noted that there are no fractions just remainders. To continue dividing the remainder the size of the counting square is increased by a factor of 10, making the counting square 100 X 100. So the numbers in the equation now become A = B/C where B = 160 and C= 3, which is A = 160/3 = 53 sets of 3 with a remainder of 1. This can continue to Infinty or One depending on the Divisor and Dividend.

The Counting table can be any 1 x 1, 10 x 10, 100 x 100, to infinity, it can also be .1 x .1, .01 x .01, .001 x .001 to infinity. When operating within a square (Counting Table) infinity is in the center of the square and infinity is also outside of the square (Counting Table).

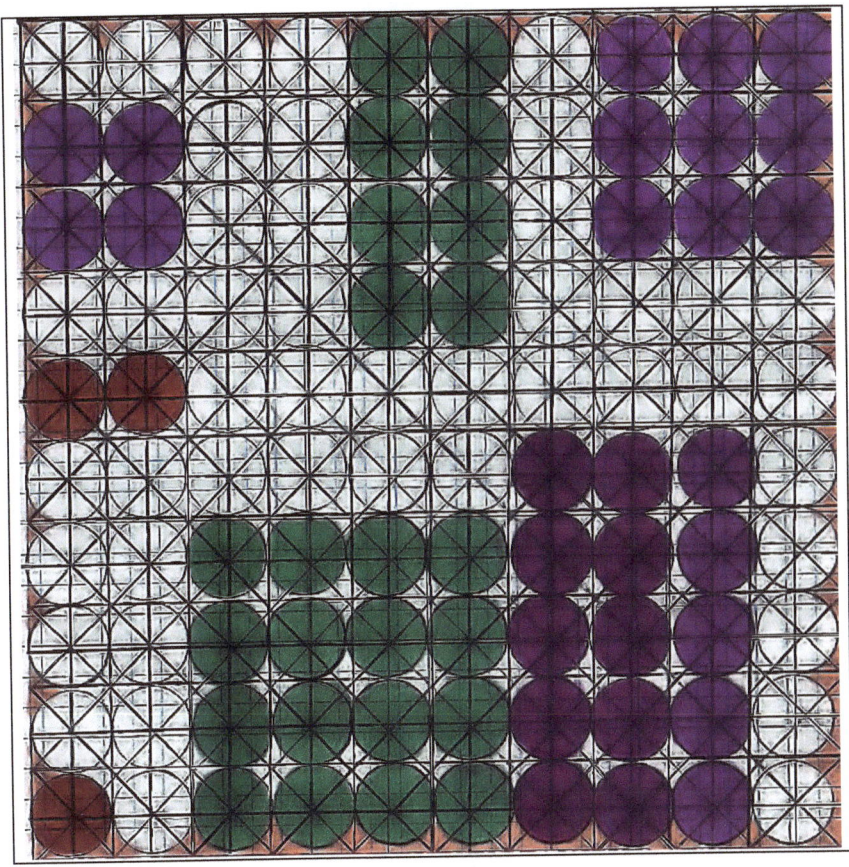

Figure 7a

Third-Generation Numbers:

The One grows to fill the square. This commences at Zero (Sunrise) and proceeds to the right and increases in size. The One does not rotate. The One square is slid and stacked. The number increases from Zero (Sunrise) until the peak of the Pyramid is reached, the number is now 2 sets of 5 vertical circles, and this continues to decrease to Zero (Sunset). All the numbers are contained within the arc that describes the motion of the Sun (Figure 8). This is then squared and is depicted in the Figure 9 construct.

Figure 8

The Figure 9 Construct has four number circles in the entire square with each having two sets of 5 Ones within a circle. Within the square are 4 quadrants. The circles are rotated creating a circle in each quadrant that is comprised of 5 single ones, horizontally and vertically. Each quadrant is 5 x 5 = 25, and 25 + 25 + 25 + 25 = 100 (Figure 10).

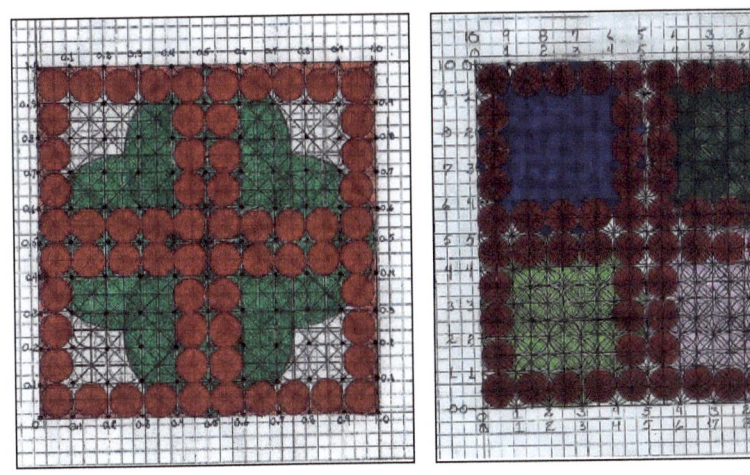

Figure 9   *Figure 10*

Fourth-Generation Numbers:

The One grew vertically and horizontally. The One also grows on the Diagonal Lines. Figure 11 is a construct of the Sun rising from the lower-left corner. It is squared on the diagonal. The root of the number is now at the intersection of the two diagonal lines (center of number) resulting in the counting of the center of the number circle.

Reference Figure 11. The One (Sun Rise) now travels up the Diagonal to the count of 1 to 5 and down to 9. The diagonal travel is from the center of one number to the center of the next number. The number lines are now in the middle of the number and are still counted as in previous constructs. The One also travels horizontally, as before, and is elevated from Zero and is the product of the Prime One.

There are several things that can be derived from this Construct. When the count is along the Horizontal it is 1 to 9. This has resulted in the numbers within the Decimal system being counted from 1 to 9, there is no Zero. The count is from center to center of the circles. They are Infinity-based Ones.

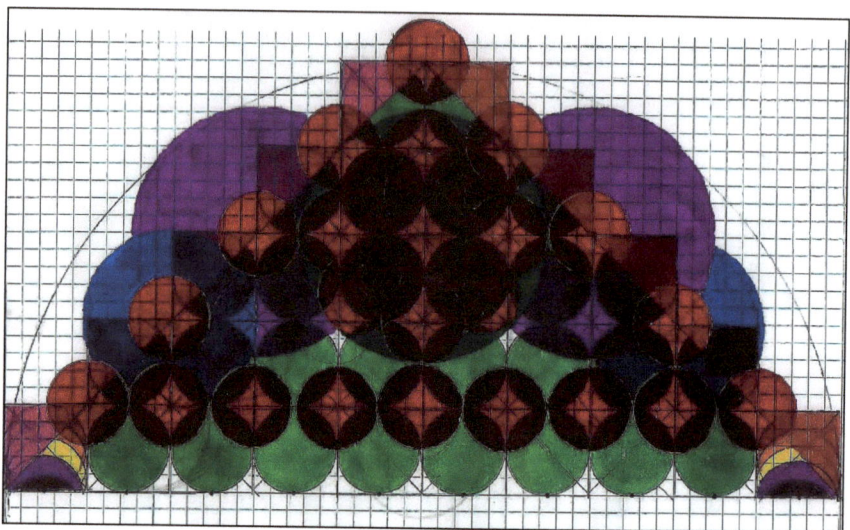

Figure 11

This construct is incorporated into a square. This is depicted in the Figure 12 construct.

4th-Generation Numbers are generated from the Infinity-based One at the center of all constructs and are combined/compared with the 2nd-Generation Zero-based numbers. And when combined the counting of Number squares can be performed.

Figure 12

This number is squared. There is an opposing diagonal count of 1, 2, 3, 5, 7. This is depicted in Figure 13.

Figure 13

When Figure 10 and Figure 13 are combined a construct as depicted in Figure 14 may be made.

Figure 14

Chapter 2
Arithmetic

Arithmetic is how Numbers are used. The symbols and words of an Equation are the language of the Arithmetic. For example: Square Root can mean Divide or Subtract. Squared can mean Multiplied or Added. If a square is made by x multiplication it means a Zero-based number has been multiplied by itself. If an infinity-based number is multiplied (•) it means the size of the number has increased Horizontally, Vertically, and Diagonally.

Arithmetical Function Symbols:

Addition:
Addition with a Zero-Based One is comprised of vertical/horizontal alignment or stacking, and the movement is left to right and or upward.

Subtraction:
Addition must occur before Subtraction; the direction of movement is from right to left and/or downward.

Multiplication:
Multiplication of numbers consist of counted groups of numbers.

Division:
Multiplication must occur before Division. Division moves toward One. The movement results in a lesser Number (Quotient). There may be remaining numbers, there are no Fractions outside the One.

Whole Numbers:

Whole Numbers consist of other proportional whole numbers (circles), to infinity. Whole numbers cannot be fractured.

Numbers and Symbols:

Numbers:

There are various forms of numbers used in Basic Mathematical calculations. There are Zero-Based numbers (N_Z), Infinity-Based numbers (N_I), Pythagorean Numbers (N_P and PN), Prime Numbers (N_{PP}), Multipliers/Divisors (N), and combinations thereof.

Symbols:

The symbols used in Basic Mathematical calculations are:

$+$ = Plus, or addition, this is movement to the right and up or up and to the right.

$-$ = Minus, or Subtraction, is a movement to the left and down or down and to the left.

X = Multiplication of a Zero-based number (N_Z or N_{ZI}).

— or / = Division of a Zero-Based number (N_Z or N_{ZI}).

$\bullet$ = Multiplication of an Infinity-Based number (N_I or N_{ZI}).

$\div$ = Division of an Infinity-Based number (N_I or N_{ZI}).

(), [],{} = grouping of numbers or grouping of a group of numbers.

Functions:

Figure 17 depicts the horizontal addition of Zero-Based numbers. Each circle is a Zero that has been made into a number. For the purposes of this construct it is a Decimal One. The count is positional.

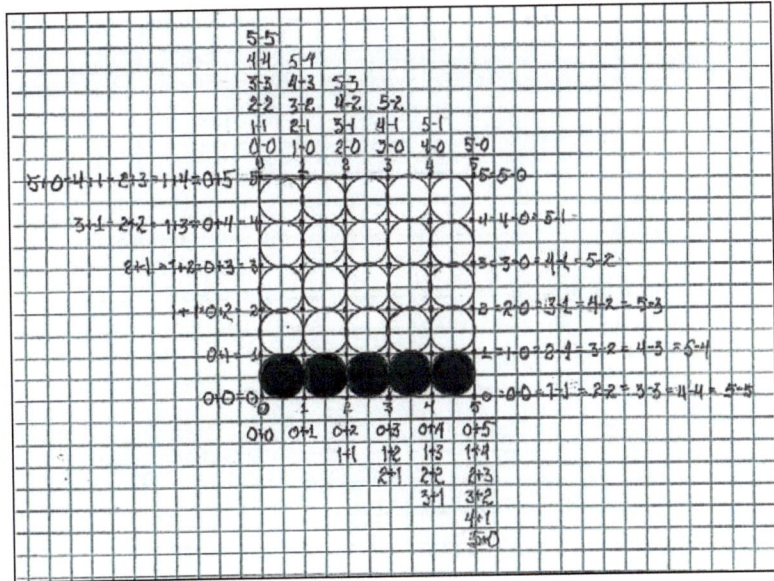

Figure 17

Figure 18 depicts the first addition, it is, 0 to 2_Z, $1_Z + 1_Z = 2_Z$ and $2_Z + 3_Z = 5_Z$.

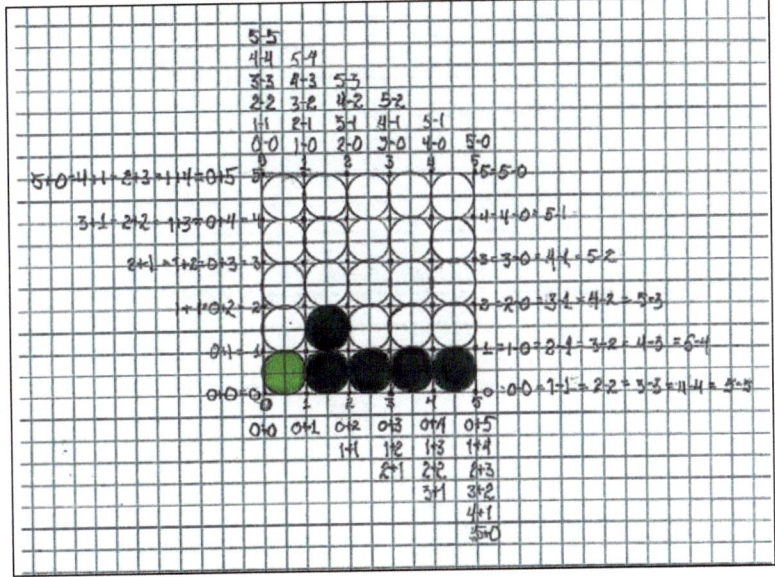

Figure 18

Figure 19 depicts 0 to 3_Z, $1_Z + 2_Z = 3_Z$, $2_Z + 1_Z = 3_Z$, and $3_Z + 2_Z = 5_Z$.

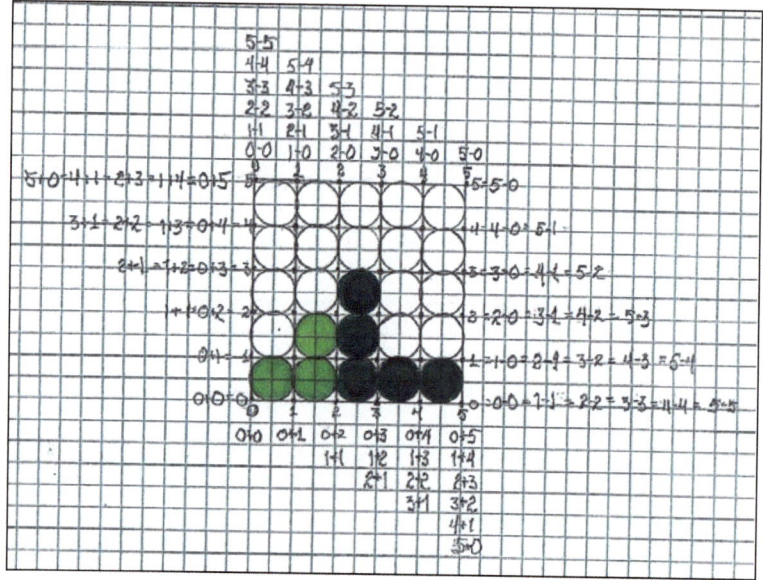

Figure 19

Figure 20 depicts 0 to 4_Z, $1_Z + 3_Z = 4_Z$, $2_Z + 2_Z = 4_Z$ and $4_Z + 1_Z = 5_Z$.

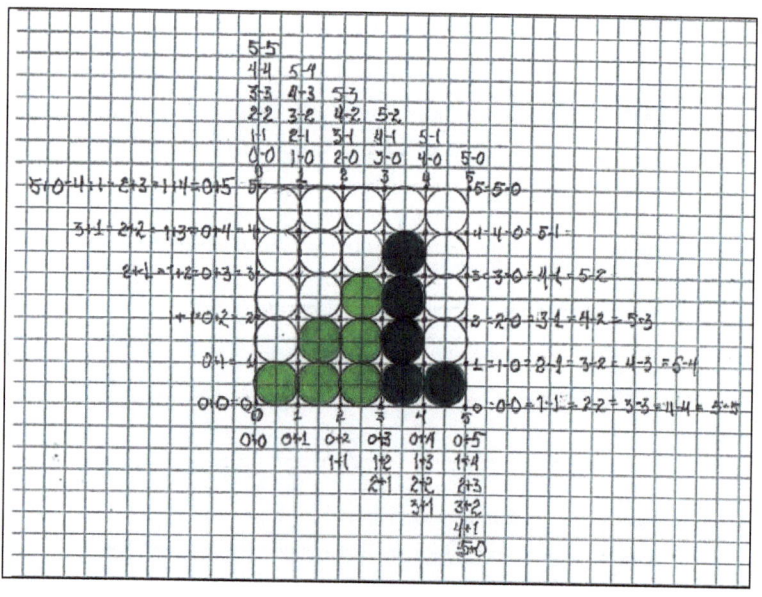

Figure 20

Figure 21 depicts $1_Z + 4_Z = 5_Z$, $2_Z + 3_Z = 5_Z$, $3_Z + 2_Z = 5_Z$, $4_Z + 1_Z = 5_Z$, $5_Z + 0 = 5_Z$.

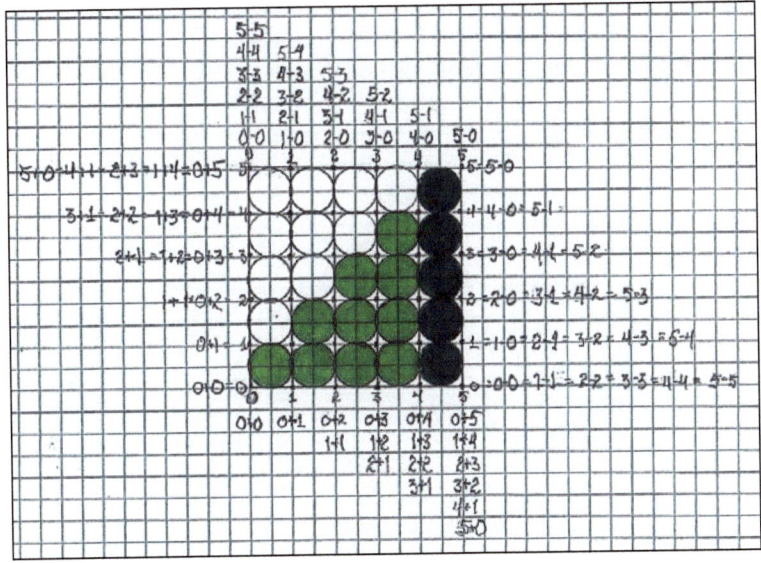

Figure 21

Figure 22 depicts Vertical addition, the count is up and to the right and is $(0 + 1_Z) + 1_Z + 1_Z + 1_Z + 1_Z$ and equals $(0 + 5_Z)$ and $1_Z + 4_Z = 5_Z$.

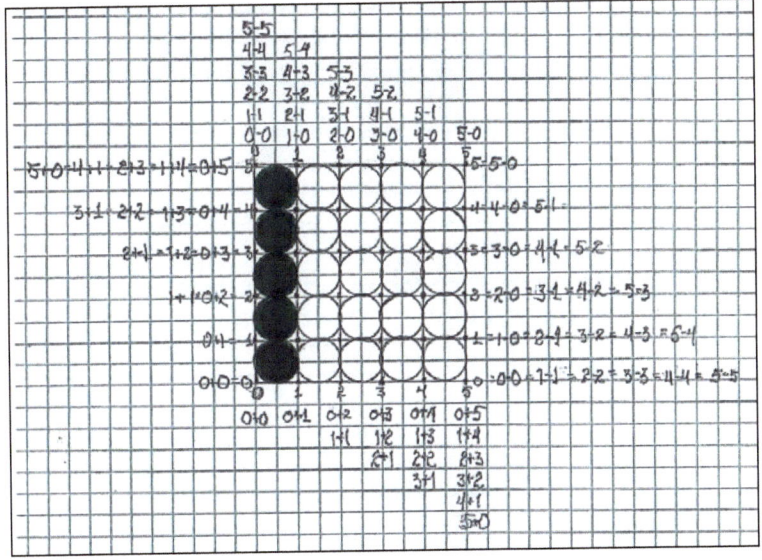

Figure 22

The Awakening of Numbers

Figure 23 depicts $0 + 2_Z$, $1_Z + 1_Z = 2_Z$ and $2_Z + 3_Z = 5_Z$.

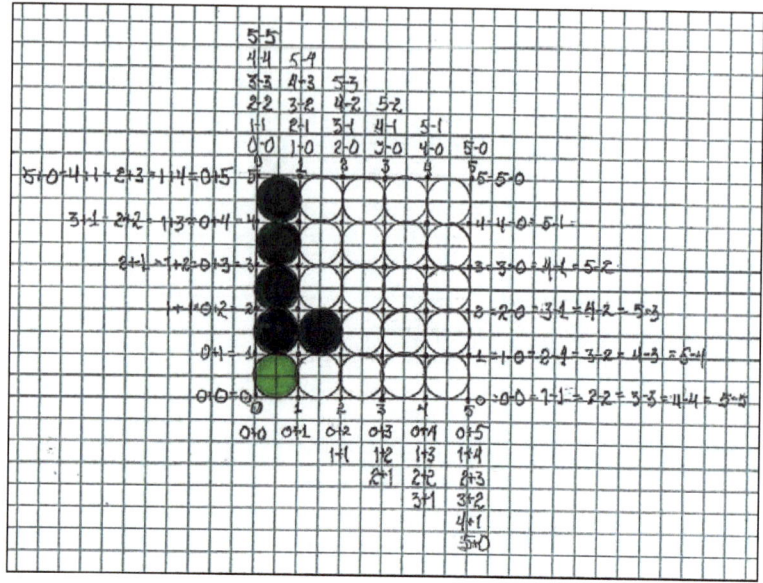

Figure 23

Figure 24 depicts $0 + 3_Z$, $1_Z + 2_Z = 3_Z$, $2_Z + 1_Z = 3_Z$, $2_Z + 3_Z = 5_Z$, and $3_Z + 2_Z = 5_Z$.

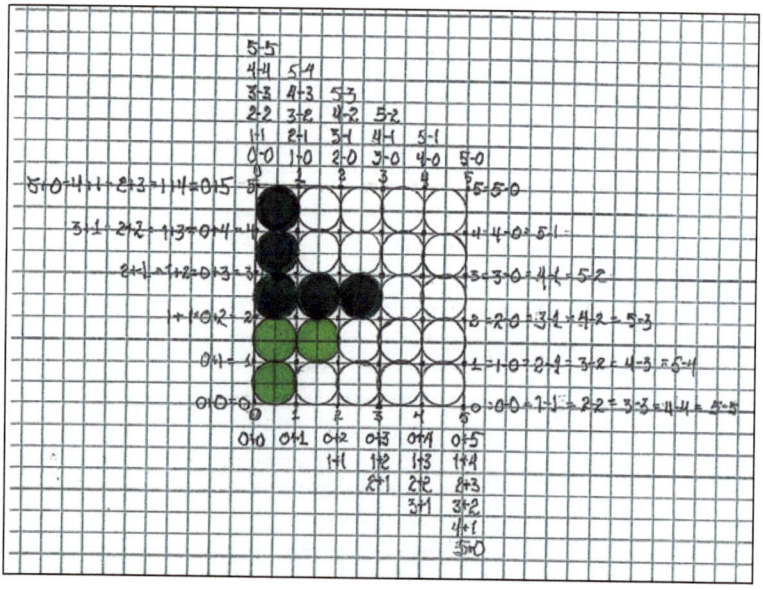

Figure 24

Figure 25 depicts $0 + 4_Z$, $1_Z + 3_Z = 4_Z$, $2_Z + 2_Z = 4_Z$, $3_Z + 1_Z = 4_Z$. $1_Z + 4_Z = 5_Z$ and $4_Z + 1_Z = 5_Z$.

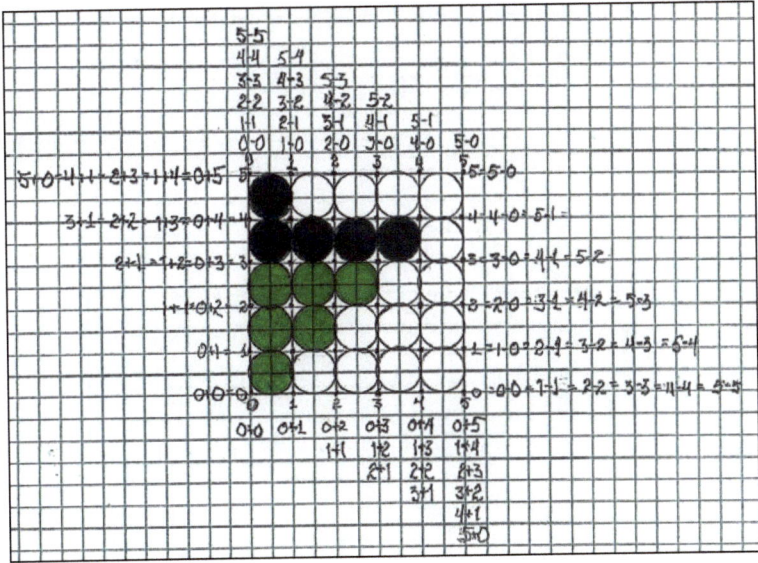

Figure 25

Figure 26 depicts $0 + 5_Z$, $1_Z + 4_Z = 5_Z$, $2_Z + 3_Z = 5_Z$, $3_Z + 2_Z = 5_Z$, $4_Z + 1_Z = 5_Z$, and $5_Z + 0$.

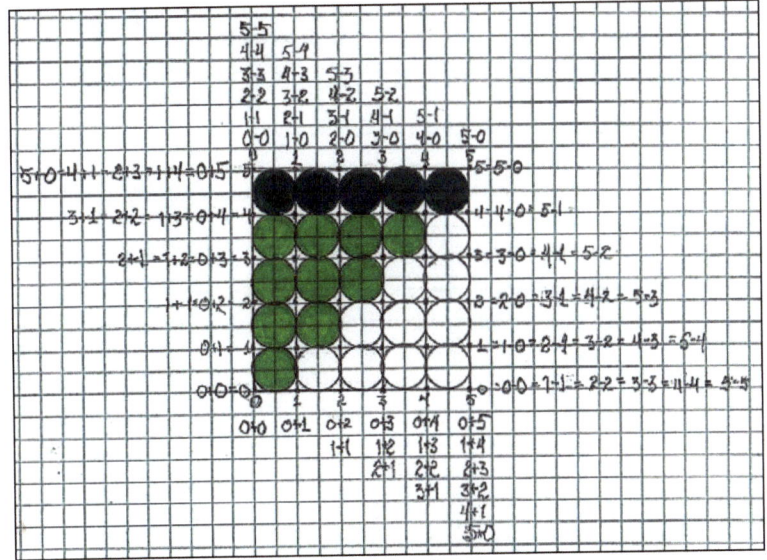

Figure 26

Figure 27 depicts the numbers created by movement of 1 to the right and up. A vertical count is applied starting with $0 + 1_Z$. A Right-Angle Triangle has been created; this becomes the horizontal base for Trigonometry. This is horizontal $1_Z + 2_Z + 3_Z + 4_Z + 5_Z = 15_Z$.

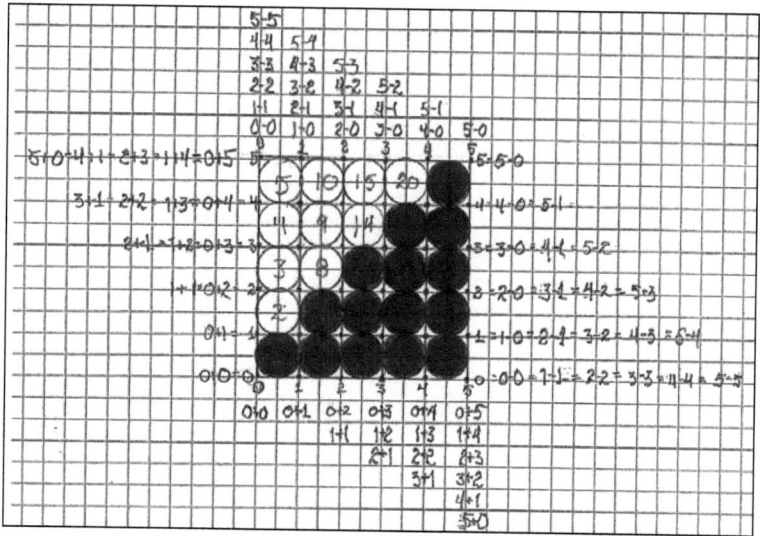

Figure 27

Figure 28 depicts the result of $3_Z + 3_Z + 3_Z + 3_Z + 3_Z$. It equals a Horizontal 15_Z.

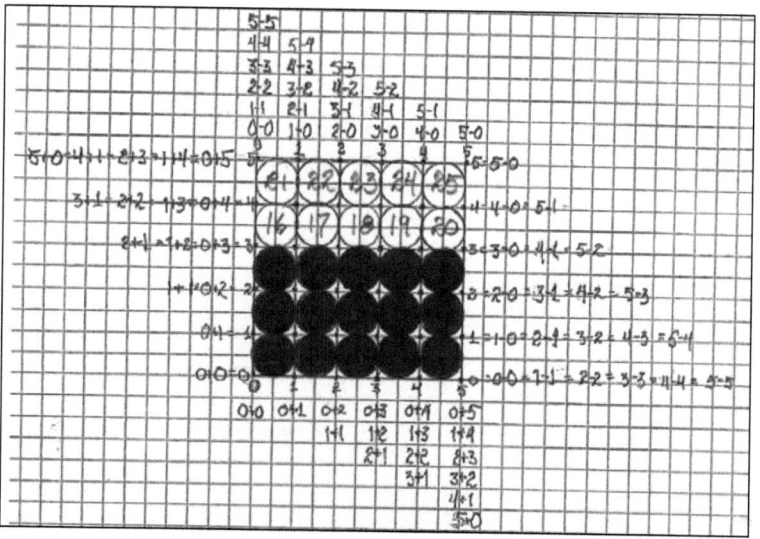

Figure 28

Figure 29 depicts the numbers created by movement of the Zero-Based 1 up and to the right. A horizontal count is applied starting with 0 + 1_Z. A Right-Angle Triangle has been created; this becomes the vertical base for Trigonometry. This is vertical $1_Z + 2_Z + 3_Z + 4_Z + 5_Z = 15_Z$.

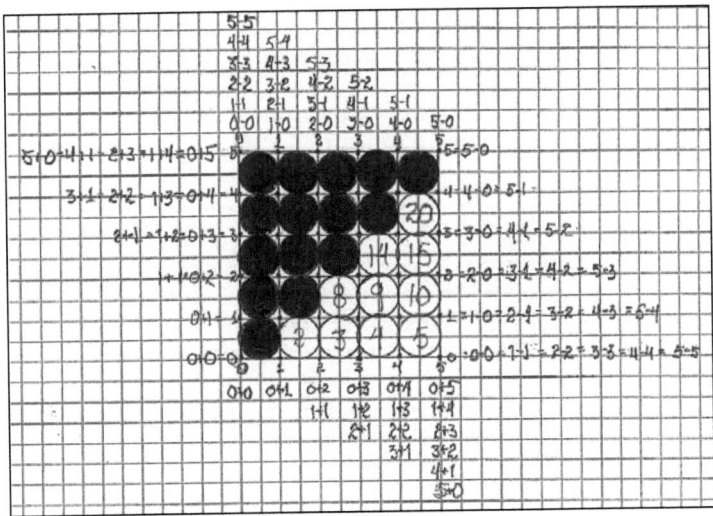

Figure 29

Figure 30 depicts the result of $3_Z + 3_Z + 3_Z + 3_Z + 3_Z$. It equals a Vertical 15_Z.

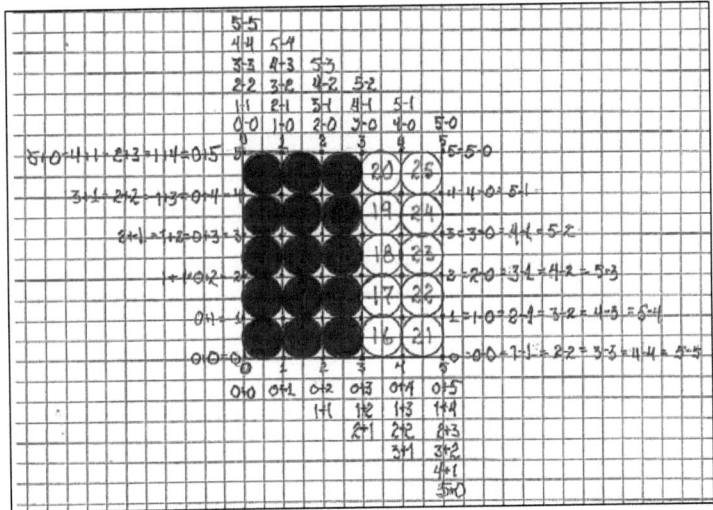

Figure 30

Squaring and Area:

The counting table is comprised entirely of Ones. There are no values assigned to the One until numbers are transferred to the Table and counted. When 10 x 10 is assigned to the Table, each square will equal 1. This sequence may continue to Infinity.

The Ones along the sides are the source of all numbers on the Counting Table, meaning One is the Root of all other numbers in this Construct and its combinations. The count begins in the lower-left corner. The first square after 1 is a positional 2 count, horizontally or vertically. A 2-by-2 square is 2 columns of 2 or 2 rows of 2, the quantifiable count is 4 squares total. This counting results in the Arithmetical function known as Multiplication and is symbolized by the X that is within the square. Multiplication is always a move toward Infinity. This multiplication continues to 10 by 10 which yields a count of 100 (Figure 31). This is squaring in appearance only. Quantifiable squaring requires the addition of diagonal Ones.

Rectangles may also be described. For example: a 2-by-3 rectangle is formed by 2 vertical squares and 3 horizontal squares. The count of the squares is 6. Or a 3-by-2 rectangle is formed 3 vertical squares and 2 horizontal squares. The count is 6.

This analysis results in the equation. Area = Length by Width.

After Multiplication counting, Division counting can occur and is symbolized by the / or —, the line that passes through the ones within a square.

Division is now possible, and the equation is: $A = C/B$, $A = 100/10 = 10$. The Division can continue to 1 and results in $1 / 1 = 1$. Division is always a move toward 1.

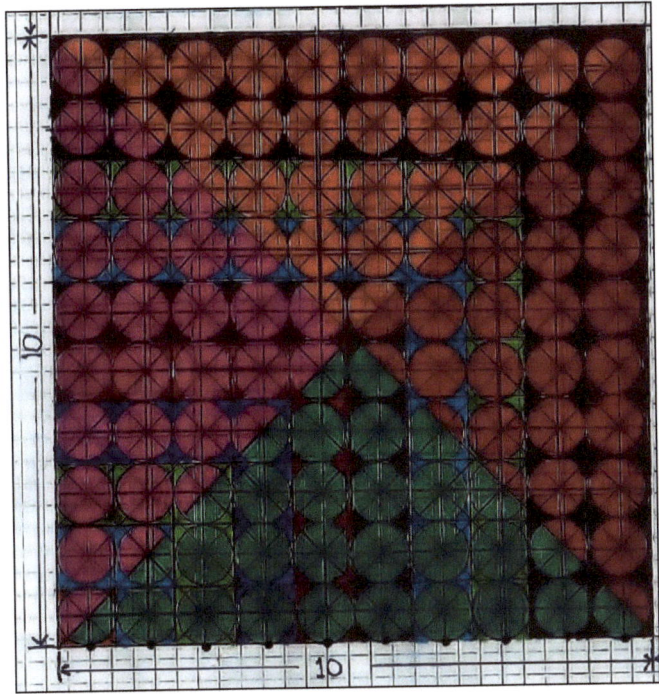

Figure 31

There are no Negative numbers, there are only Whole numbers (Figure 31). All numbers are comprised of a ones. When we multiply it is still A by (X, times) B = C. 1 X 1 = 1, 1 X 2 = 2, 2 X 2 = 4, 2 X 4 = 8, 3 X 3 = 9, 3 X 5 = 15 and 4 X 4 = 16. Division is counting the number of times a specific number is contained within the larger number, it is symbolized by / or —.

The Counting table can be 1 x 1, 10 x 10, 100 x 100, to Infinity. Infinity is outside of the square (Counting Table).

All numbers with a fraction added are multplied by multiples of 10 until the desired Whole number is attained. When all computations have been completed the resulting Whole number is divided by the number of multiples of 10 that were used to achieve a Whole number.

The squaring of the number 15_Z, that were presented in Figures 27 and 29, results in the number of 30_Z for each square. This is the squaring of 5_Z and is not the same as 5^2 or $5_Z \times 5_Z$ which results in 25_Z.

The Awakening of Numbers | 25

Figure 32 depicts the perimeter count of the Horizontal and Vertical Right-Angle triangles. The left Vertical Line and the Bottom Horizontal line will become the base of the triangles. The perimeter count is 16_Z.

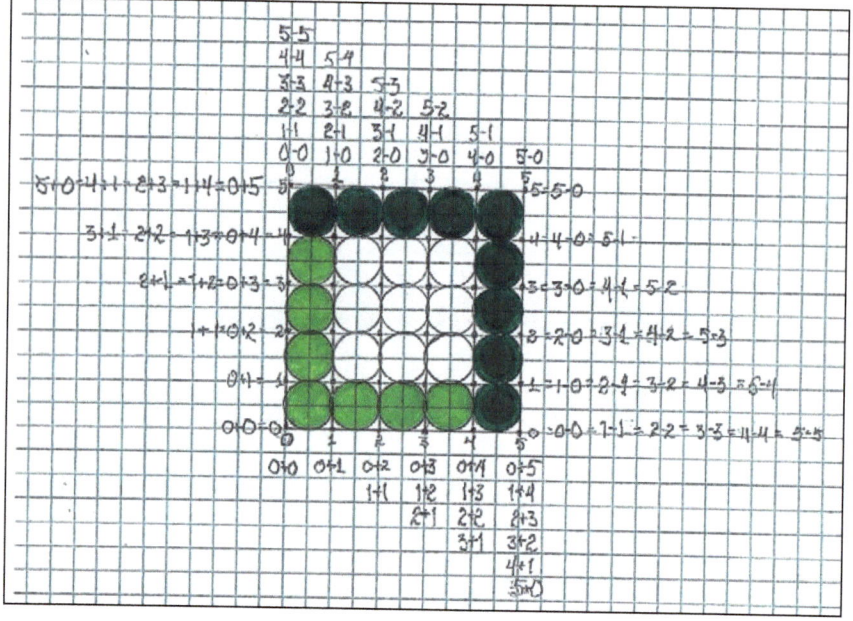

Figure 32

Rows and Columns

Zero-Based Numbers (N_Z) are limited Infinite columns or rows. The rows and columns may be limited in various ways. Thus far numbers may be created that resemble a single column or row, a combination of a column or row, a rectangle or a right-angle triangle. All columns and rows are limited to 5_Z so far. Numbers are created from the lower-left corner and progress either upward and/or to the right or to the right and/or upward.

Figure 33 depicts the addition of rows or columns.

Columns are created and continue to the right and upward creating additional columns until limited.

Rows continue upward and to the right creating additional rows until limited.

The number of 1_Z in five 5_Z rows or columns is 25_Z. Squares are formed by the rows or columns.

The count of each row and column is shown in Figure 33 that represents a complete square.

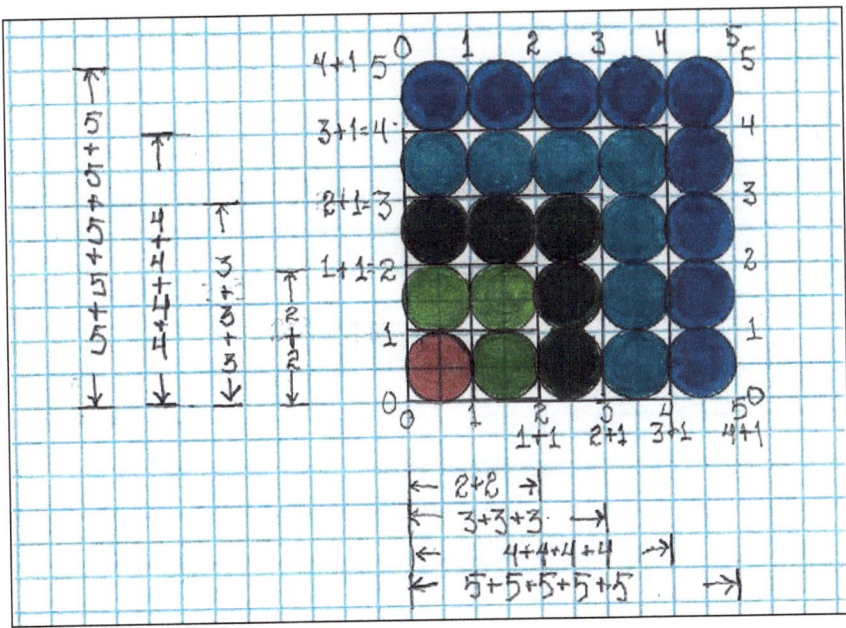

Figure 33

Subtraction

Numbers must be created (Added) before Subtraction occurs. A larger number (Subtrahend) cannot be subtracted from a smaller number (Minuend). The smaller number (Minuend) must be increased to a number in multiples of 10 until the number is larger than the Subtrahend. The resulting answer is then divided by the number of 10 multiples. Example: 11/20 = ?. 11 x 10 = 110, 110/20 = 5, 5/10 =.5 + remainder of 10, 10 x 10/20 =5, 5/100 =.05, .5 +.05 = .55.

Decimal Positional Notation

Zero is not a digit; it is used as a place holder in Positional Notation. The nature of Zero is the absence of numbers. The absence of a number can and is represented by numerous symbols, including a blank.

Only one decimal digit is contained within one position/space and it is expressed with a decimal number symbol of 1 to 9 in addition to the symbol for the absence of a digit.

Positional Notation is a visual display only. The format currently reflects decimal number symbols and formatting. Mathematical calculations and counting are performed prior to being displayed as Decimal Positional Notation, regardless of the mathematical methodology being used for the relevant Base-numbering system.

The format of the Decimal Positional Notation indicates the count is within a circle that is not externally limited, and is as follows:

$\infty \leftrightarrow 0$

$10^\infty \leftrightarrow 10^{10} \times N + 10^9 \times N + 10^8 \times N + 10^7 \times N + 10^6 \times N + 10^5 \times N + 10^4 \times N + 10^3 \times N + 10^2 \times N + 10^1 \times N + 10^0 \times N$

Infinity $\leftrightarrow$ 10 billion x digit + 1 billion x digit + 100 million x digit + 10 million x digit + 1 million x digit +

100 thousand x digit + 10 thousand x digit + 1 thousand x digit + 1 hundred x digit + ten x digit + one x digit

The count is from the center (∞) outward to the limit of Zero (No Numbers).

The results of a fractured One, also known as a Fraction, are also used in Decimal Positional Notation. A Decimal Point (.) is placed to the right of the displayed Whole Number and is added to the Whole Number Positional Notation display. This effectively means a decimal point (.) serves as a symbol for addition (+). The formatting is as follows:

$1 \leftrightarrow \infty$

$N.10^{-1} \times N + 10^{-2} \times N + 10^{-3} \times N + 10^{-4} \times N + 10^{-5} \times N +$

$10^{-6} \times N + 10^{-7} \times N + 10^{-8} \times N + 10^{-9} \times N + 10^{-10} \times N \leftrightarrow 10^{\infty}$

Whole Number + (.) 1/ten x digit + 1/1 hundred x digit + 1/1 thousand x digit + 1/10 thousand x digit + 1/100 thousand x digit + 1/1 million x digit + 1/10 million x digit + 1/100 million x digit + 1/1 billion x digit + 1/10 billion x digit ↔ 1/∞

Chapter 3
Numbers

Zero-Based Numbers:
The absence of a Number is referred to by various names and symbols whose attributes are imposed upon the absence. For the purposes of this discussion the English word Zero and the symbol O will be used. Understanding Zero means the absence of a number, not the absence of everything.

As with numbers Zero is imaginary and is visualized as a circle (O). The circle contains no numbers within the circle and there are no numbers external to the circle (Reference Figure 34). The distance around the circle is infinite. When the circle is viewed edge-on and twisted the symbol ∞ (infinity) is formed.

When the circle is visualized as containing numbers, the numbers are within the circle and are infinite (∞). The diameters and radii within the circle are also ∞. There are no numbers external to the circle (Reference Figure 34).

When a circle is visualized as a quantifiable number the distance around the circle is known as the circumference it is measured in degrees. Within the circumference are quantifiable diameters and radii.

The vertical and horizontal diameters are repeated external to the number. These lines are quantified. This results in the corners of the square being designated as a Zero (Reference Figure 34). For the Decimal system the count is limited from 0 to 10 along each line, or multiples of 10, i.e.: 10 x 10 x 10 x 10 ⟶ ∞.

Even though Zero-based numbers are shown as a circle, they are not. They are 1 square with one circle. This results in a zero-based number square. Zero-based numbers cannot rotate. They are designated as N_Z.

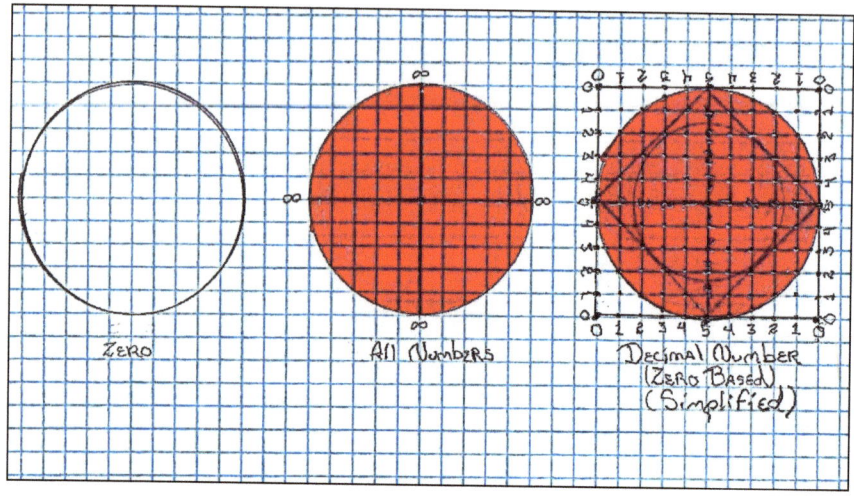

Figure 34

All Zero-Based (N_Z) numbers are a result of counting and/or adding of 1_Z. Numbers must be created (Added) before Subtraction can occur. Zero-Based numbers (N_Z) extend to Infinity.

Lines:

Lines are used to define numbers. When a square is made it gives number diameters, without lines a number is a circle, which is infinity in circumference. When a number is squared all corresponding numbers are also squared, which results in the diameters squaring an adjacent number. All lines are shared by all numbers. Lines and the intersection of 2 or more lines are given names. The square and the circles (numbers) enclosed in the square define the limits of the infinite lines. The names of the lines used for a Mathematical procedure should be used only when performing analysis with the equations relevant to the Mathematical system being used.

There are numerous names for lines, many only serve to make Mathematical riddles and serve no function in the understanding of numbers. We will confine our analysis of numbers only using the names that are relevant to the Arithmetic being used.

The Zero-Number Armature:

The Zero-Number Armature shown in Figure 35 applies to all numbers in the Decimal system, regardless of the size of the number (circle).

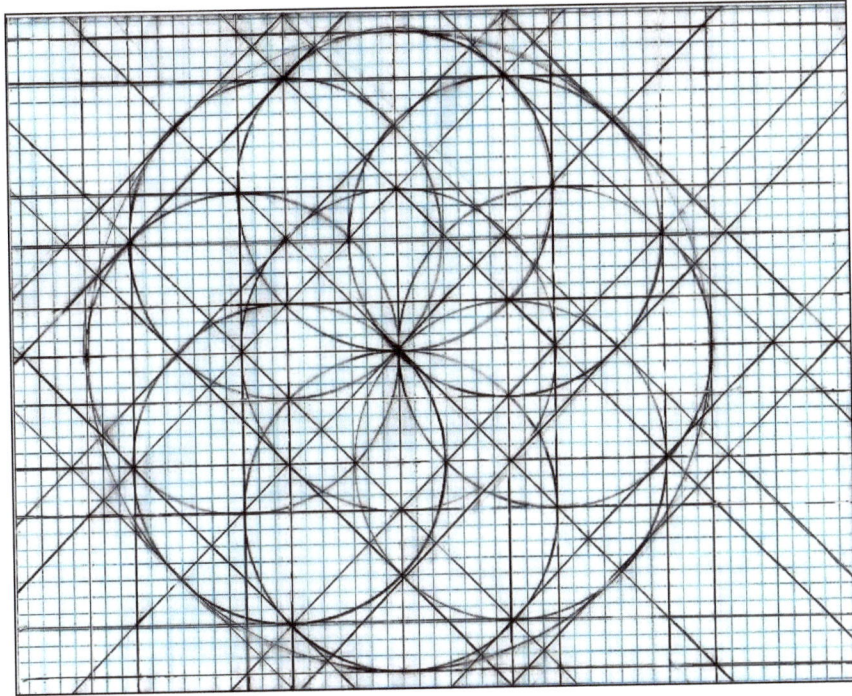

Figure 35

Infinity-Based Numbers:

The center of a number is infinite as is the circumference until limited. The count starts in the center and expands outward. The count for Addition, Subtraction, Multiplication and Division are along all Radii, simultaneously. The count expands outward until limited by a square. This results in Numbers that are externally squared. Within the squared Infinity-Based number are Radii, Diameters, and a diagonal square. The diagonal square within the circle may rotate. This is the rotation of numbers within the number, the externally squared number cannot rotate.

The count of Infinity-based numbers is from center to center and begins at 1. Since there is no Zero, there is no 10, the count is 11. Infinity-Based numbers are designated as N_I.

Complex Numbers:

Complex Numbers are a combination of a Zero-based number (N_Z) and an Infinity-based number (N_I). They are designated as N_{ZI} and are complete squares. The squares may rotate and contain whole numbers and fractions or are a fraction.

Decimal Numbers:

The Prime One is the source of all numbers in the Decimal system. The Prime One generates Zero-based numbers, the Infinity-based numbers, Infinity-based numbers squared, the Pythagorean Numbers and Pythagorean Numbers squared.

All Numbers initially have no dimensions; they are without diameters until they are squared internally or externally. When a number is put on the counting table it is assigned a dimension. For the Decimal System a Square (Counting Board) is assigned 10 x 10 divisions, and each of these divisions are 10 x 10 divisions and this continues to Infinity. When the size of number is determined, the number is then assigned to a Counting Board.

Pythagorean numbers (10p) use Decimal number symbols, Arithmetical functions are performed the same for 10p numbers as they are for Decimal numbers.

Figure 36 shows a Number superimposed on the Basic Counting Board. If the corners are designated as Zero, then Addition (+) and Subtraction (-) may be performed to Zero. Multiplication (x) and Division (— or /), may be performed to 1. If the center is designated as Infinity, then Addition (+) and Subtraction (-) can only be performed to 1. Multiplication ( ) and Division (÷) can also only be performed to 1.

Figure 36

Pythagorean Numbers:

Pythagorean Numbers are incorporated within the Decimal System. They are 4th-Generation numbers. The incorporation of Pythagorean Numbers within the Decimal system will be referred to as Np or PN. They are depicted in Figure 37.

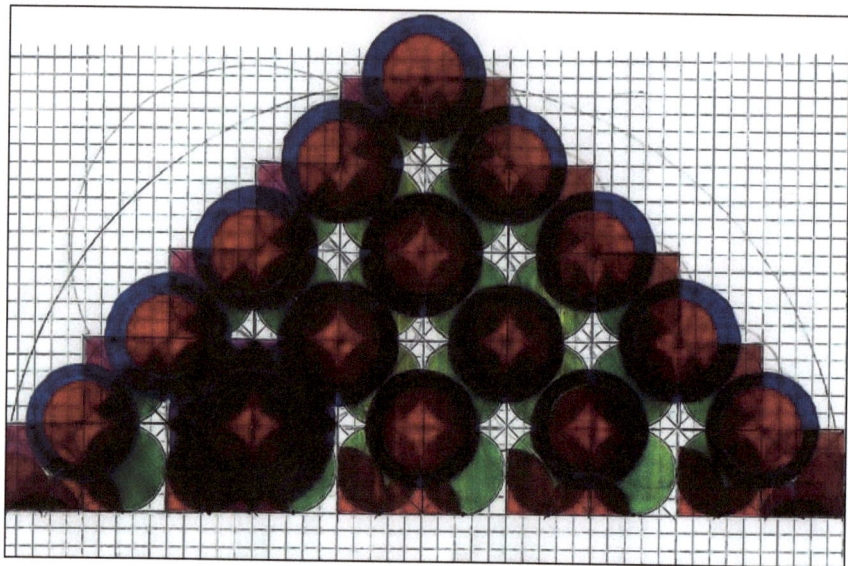

Figure 37

In the lower left of this construct is the Prime Number One and is shown being formed from the corner. After the center of One has reached the first intersect of the vertical and horizontal one line it is squared, and a circle is then drawn around the square. The same procedure is performed with all the remaining diagonal Pythagorean Ones in the Decimal System.

Figure 37 shows the Pythagorean Numbers superimposed over the Second-Generation Decimal Number Construct. Decimal number symbols correlate with the Pythagorean Numbers.

Figure 38

Figure 38 depicts a Decimal 4-quadrant system construct. On the diagonals, the Pythagorean Decimal Count is 1p, 2p, 3p, 4p, 5p. From the peak of each quadrant down is 5p, 4p, 3p, 2p, 1p. The opposing diagonal count are a Prime Number count. The single-digit Prime Number is a count of 1pp, 2pp, 3pp, 5pp, 7pp. These are single-digit Prime numbers; it is the source for all Prime numbers. The next number is the 2-digit number 11pp. The count continues in accordance with the Prime Number counting Rule: A counted Prime Number can only be divided by itself and result in 1. Reference Figure 39.

Figure 39

The Pythagorean Equation ($A^2 + B^2 = C^2$):

The Pythagorean equation is commonly referred as the Pythagorean Theorem because of the misapplication of Arithmetical functions and misinterpretation of what was being said. What was said is: When working within a square A and B are the same as C. This is demonstrated in the Figure 40 construct. Pick any 3 adjacent number Radii, within the number circle, and designate them as $\sqrt{A^2}$, $\sqrt{B^2}$ and $\sqrt{C^2}$, you will see they are all equal. And you will see they are all equal down to the center one. What this means is, while rotating any radius, and holding C constant, a circle will be formed within the square. This is the beginning of Geometry.

Figure 40

The misapplication of the of the Pythagorean equation Arithmetical functions is a result of the statement of the equation and then presuming a different equation statement. This results in not Multiplying and Dividing, before Adding and Subtracting. The equation as stated is $A^2 + B^2 = C^2$ this means, solve for C^2. If it is desired to solve for C, then the equation becomes: $\sqrt{A^2} + \sqrt{B^2} = \sqrt{C^2}$, or $A + B = C$.

$A^2 + B^2 = C^2$ is the measurement of a circle with 2 squares enclosed, one vertical and horizontal and a second diagonal square at 90 intervals. This configuration is considered Pythagorean (N_P) and is shown in Figure 41.

Figure 41

Chapter 4
Circular Counting

Prime One Expansion

The Infinitely small Prime One (•) is expanded as shown in Figure 42. The resulting Circle is externally squared resulting in 2 measurable diagonal lines from corner to corner. The resulting square is then circled and a measurable vertical and horizontal are created and diagonally squared. This expanded measurable Prime one becomes the measuring circle for all further expansions. Polarity of count is now determined.

Counting along the horizontal and vertical lines results in the following. The count in the upper-right quadrant is vertically up and horizontally to the right, + and +. The count in the upper-left quadrant is vertically down and horizontally from left to the right, - and +. The count in the lower-left quadrant is vertically down and horizontally to the left, - and -. The count in the lower-right quadrant is vertically up and horizontally to the left, + and -.

Figure 42

Move 1.

Figure 43 shows the results of the first movement of the right horizontal line to the right vertical line.

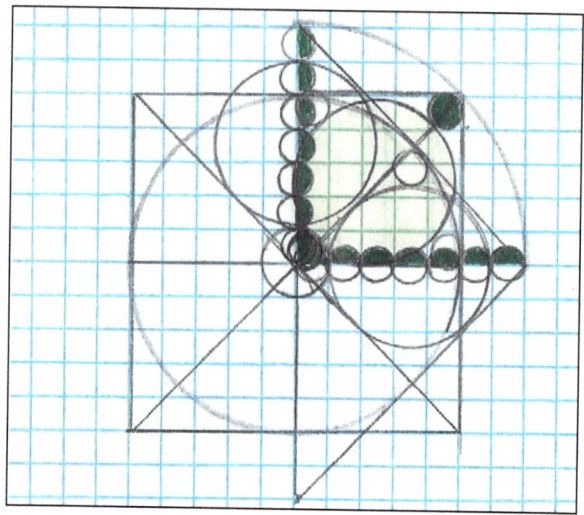

Figure 43

This movement results in the generation of a Zero-Based Number in the upper-right quadrant. The corresponding Decimal Zero-Based Counting Board quadrant is shown in Figure 44.

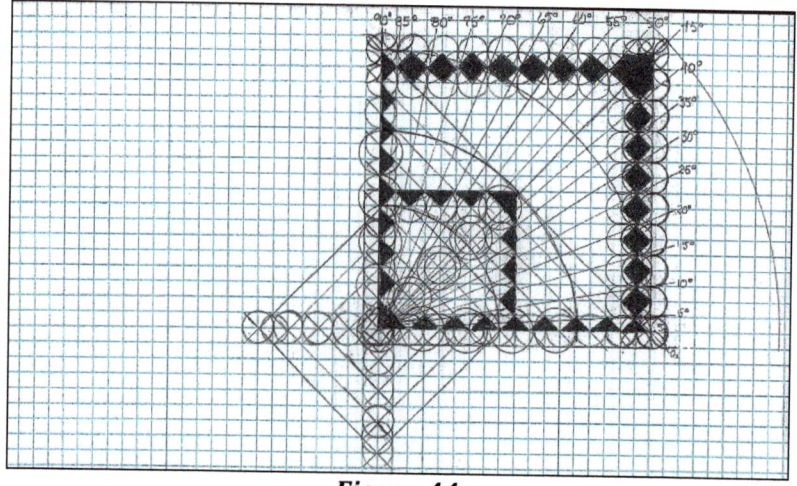

Figure 44

The placement of the Decimal Counting Board results in the Prime One radius line extending out to middle of 10_Z ($.5_Z$). The first vertical move as depicted in Figure 44 is 5 . The count continues in a counterclockwise direction around the outer perimeter of both depicted squares until the line is vertical. This results in two 90° squares as shown in Figure 44.

Move 2

The counterclockwise rotation from vertical to horizontal, within the Prime One, continues as depicted in Figure 45. The rotation is Infinity based and a Zero-Based number now exists in the upper-right and upper-left quadrants.

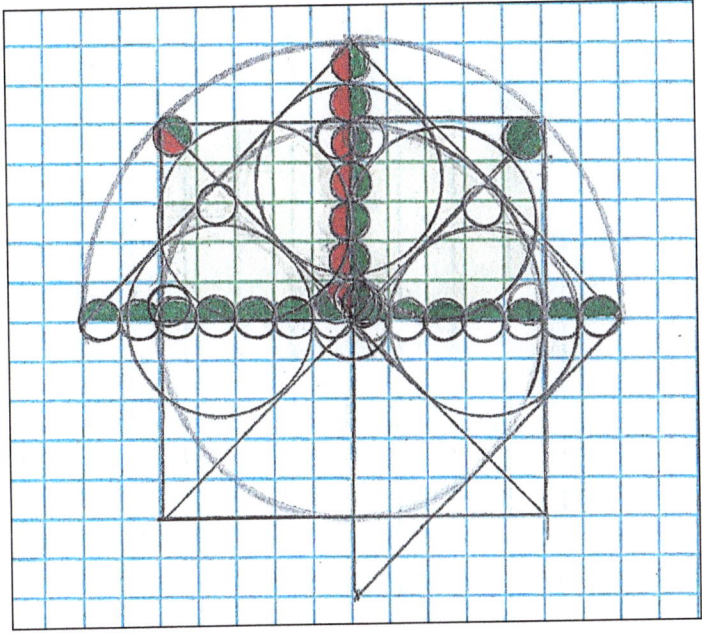

Figure 45

The Decimal Counting Board is moved (slid) to the left creating the upper-left quadrant as shown in Figure 46.

The Awakening of Numbers | 43

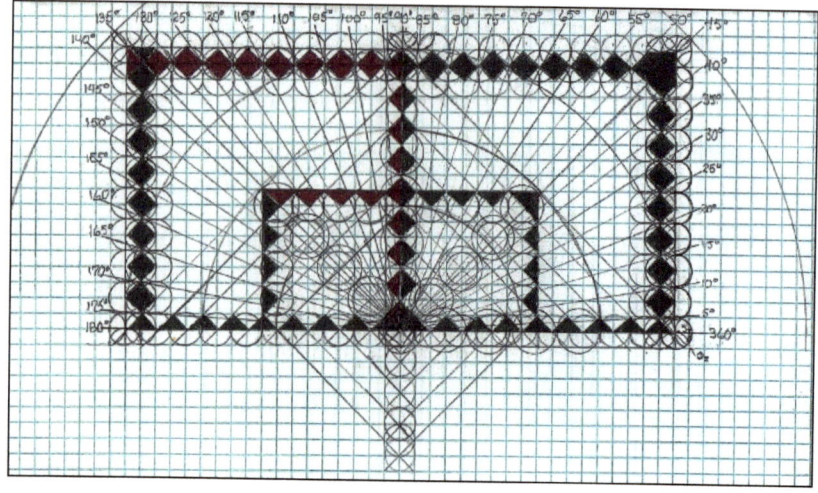

Figure 46

The count continues from the vertical 90° line to the horizontal 180° as shown in Figure 46.

Move 3

The counterclockwise rotation continues within the Prime One resulting in 3 quadrants containing Zero-based numbers that are created by the rotation of an Infinity-based number, as depicted in Figure 47.

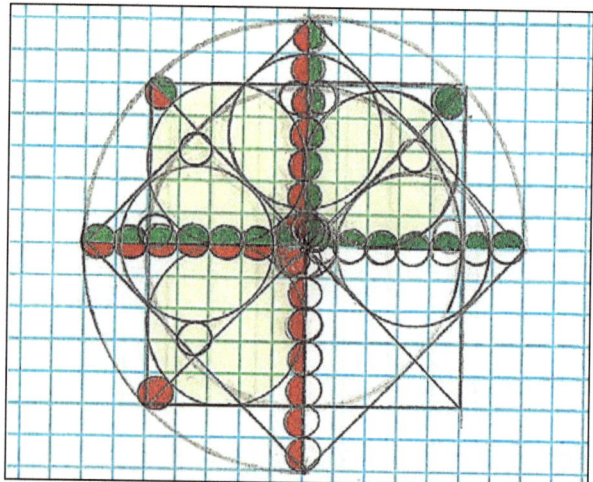

Figure 47

The Decimal Counting Board is moved (slid) down from the upper-left quadrant creating the lower-left quadrant as depicted in Figure 48.

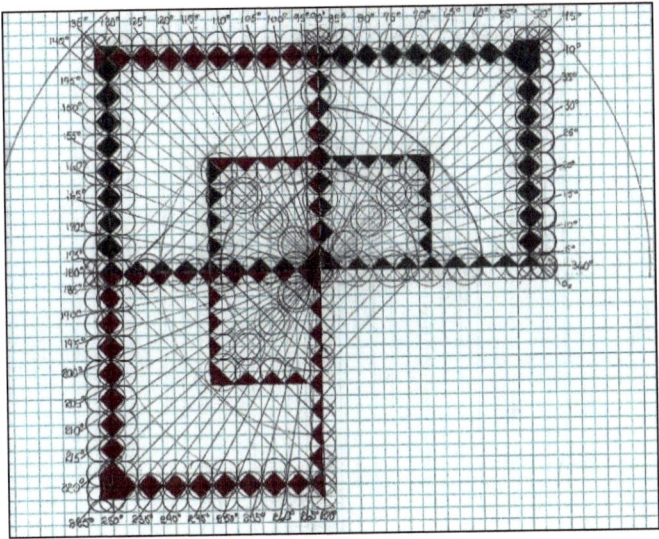

Figure 48

The count continues counterclockwise from the left horizontal 180° line to the downward vertical line reaching 270° as shown in Figure 48.

Move 4

The counterclockwise rotation within the Prime one continues resulting in 4 quadrants with a Zero-based number as a result of the rotation of an Infinity-Based number as depicted in Figure 49.

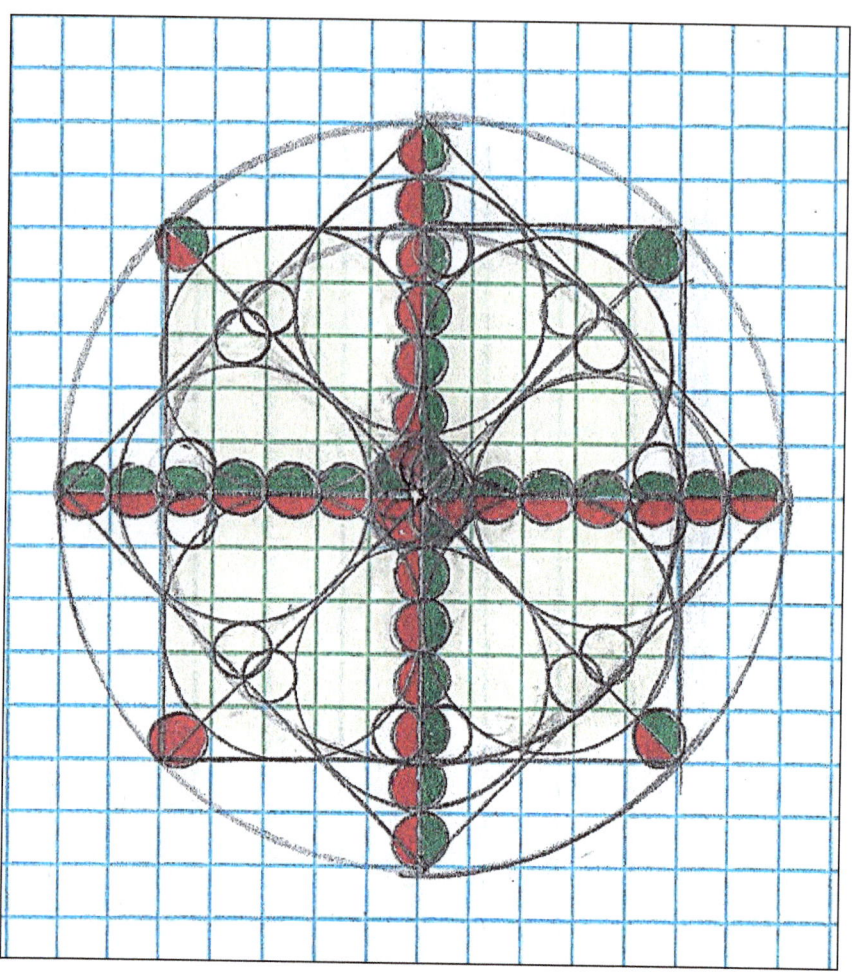

Figure 49

The Decimal Counting Board is moved to the right (slid) and Whole numbers are superimposed as depicted in Figure 50.

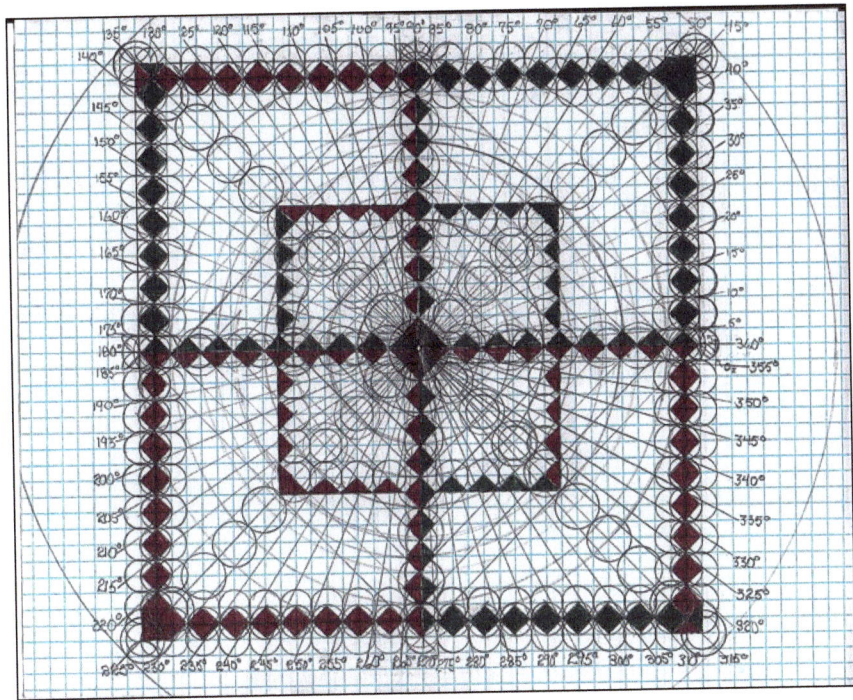

Figure 50

    This completes the making of the initial 360° Circle. This initial Circle is the source of a Year being 365 and ¼ days, with the distance between 0_Z and $.5_Z$ being the ¼ day.

    All Circumferences of all circles are 360° after creation of the initial circle. The circumference of all circles is Infinite until limited by lines.

    Degrees are an independent Number system and may be compared and measured in conjunction with other numbering systems. Typically, this is done with Np_{ZI} and the appropriate annotation is N.

    Upon completion of the initial Rotation the limiting lines have several designations. The initial line extending from the Prime One may be referred to as Adjacent or Radius/Hypotenuse when rotated. When rotated it results in 4 shared internal lines extending outward from the Prime One vertically and horizontally. When combined 2 Radii/Hypotenuses become one Diameter and the Diameter may rotate within the entire square. The rotation also results in diagonal lines that are shared among the right-angle triangles and may be designated as Radii/Hypotenuse and when combined are also considered Diameters and may rotate.

There are 4 quadrants consisting of 2 right-angle triangles in each quadrant. Upon completion of the initial rotation there is 1 vertical and 1 horizontal perimeter line in each quadrant.

When the initial line is designated as Adjacent it will not rotate. It is a shared quadrant perimeter line. The Adjacent line may be Horizontal or Vertical depending on the right-angle triangle that is being evaluated. Upon designation of the initial Horizontal line as Adjacent the adjoining Vertical line is designated as Opposite and is also a shared perimeter line that may be vertical or horizontal depending on the right-angle triangle being evaluated. This results in the creation of the equation a + b = c, with a = Adjacent, b = Opposite and c = Hypotenuse.

When the Radius/Hypotenuse/Diameter is rotated and when measured with the perimeter lines the term for the combined Adjacent/Opposite measurement lines is Tangent. This combination results in the equation of Tangent = Opposite ÷ Adjacent. This results in the perimeter lines being treated as a single line and can result in the equation solution of ∞.

Within the Prime One the vertical and horizontal lines are designated as Sine or Cosine. The equation for these lines is: Sine = Opposite ÷ Hypotenuse and Cosine = Adjacent ÷ Hypotenuse. The Sine or Cosine can never be greater than One.

Chapter 5
The Making of the Square Root of Two

The Square Root of Two is created within the Prime One. All creation operations performed with the Prime One can be increased by multiples of 10 to Infinity. The Zero armature for development of the Square Root of Two is depicted in Figure 51.

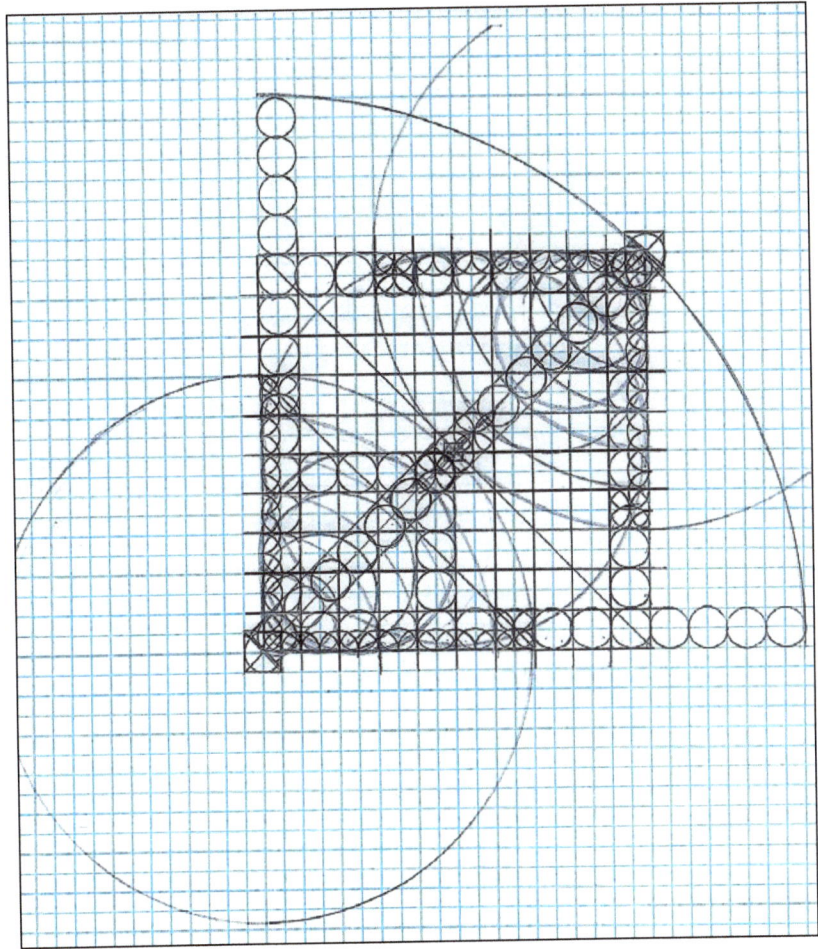

Figure 51

The diagonal numbers are referred to as Pythagorean Numbers. When rotated to the Horizontal or vertical they may be referred to as the Hypotenuse. The rotation of the diagonal to the vertical and horizontal is shown in Figure 52. The diagonal is a Pythagorean 1 comprised of 10 x $.1_p$ that becomes a 1P Hypotenuse that is rotated to the diagonal from the horizontal and vertical positions.

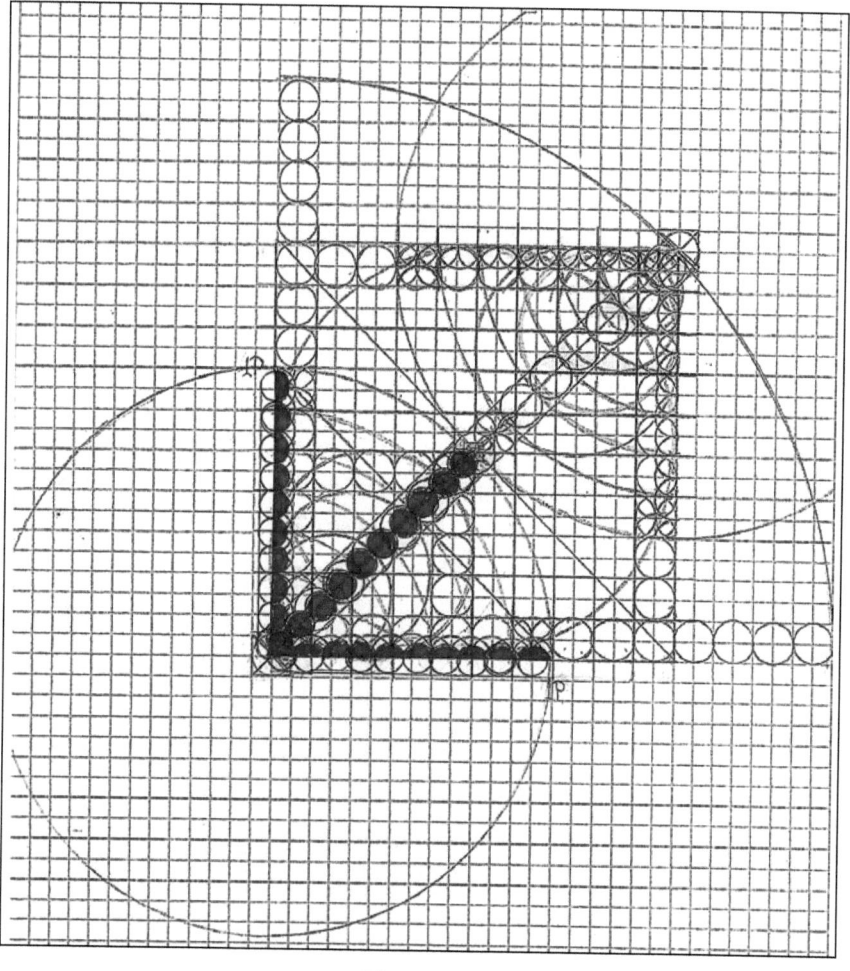

Figure 52

After the N_P numbers are rotated to the horizontal and vertical positions, they are incorporated within the N_{ZI} numbers. The movement to each intersection of infinite lines, within the Prime One, is counted horizontally and vertically as shown in Figure 53. The line intersections are defined as "Bits." The count is converted to Binary Coded Decimal (BCD), this is a squared number count as shown in Figure 53. The squared "Bit" count results in the current English Electronic Computer terms of 1 = 1 = Bit, 01 = 2 = Binary, 001 = 4 = Nibble, 0001 = 8 = Byte, 00001 = 16 = Word.

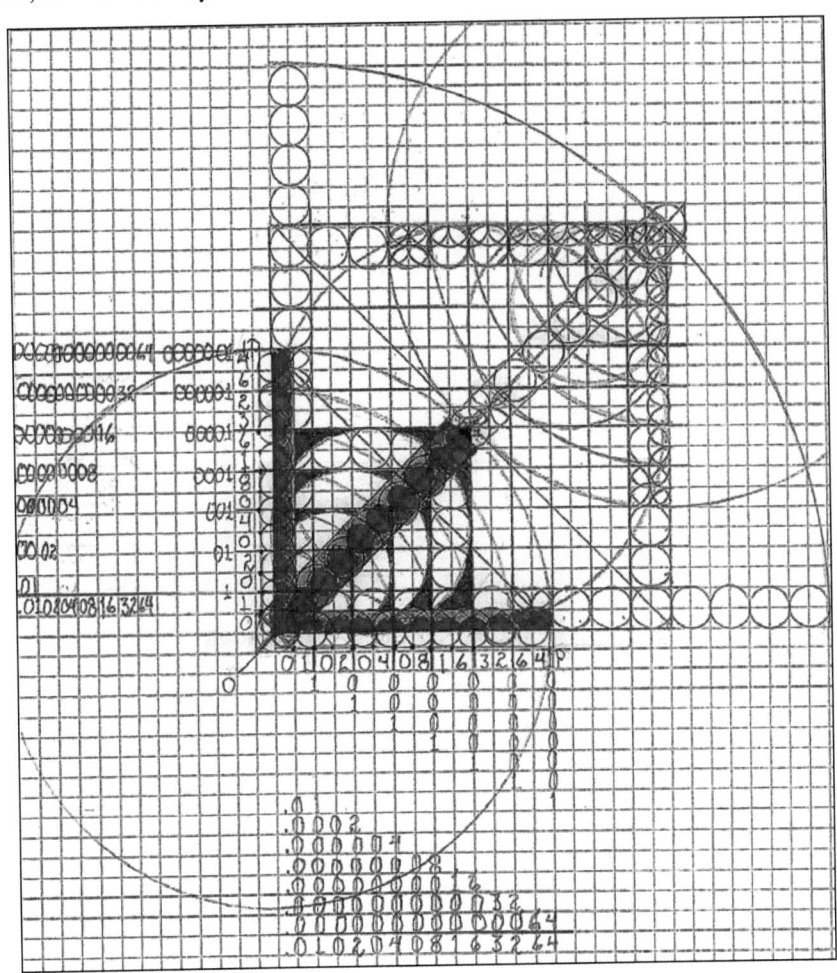

Figure 53

The 4-quadrant square count is 10 horizontal squares x 10 vertical squares or 10 vertical squares x 10 horizontal squares, yielding a larger square count of 100. The squared BCD number, within the Prime One, is .0102040816 as shown. The $\sqrt{.0102040816}$ is .101015254 and when multiplied by 7 results in a Hypotenuse count of .70710678. When the Hypotenuse is rotated to the diagonal, the $.70710678_{ZI}$ number count is now contained within a $.5_{ZI} \times .5_{ZI}$ square.

The entire 4-quadrant square is rotated 180 and the previous count is applied to the lower-left quadrant $.5_{ZI} \times .5_{ZI}$ square, as shown in Figure 54. This results in a count of two $.70710678_{ZI}$ Hypotenuse contained within a $10_{ZI} \times 10_{ZI}$, 4-Quadrant, square. Which is a 1.414213562_{ZI} diagonal count as shown in Figure 54.

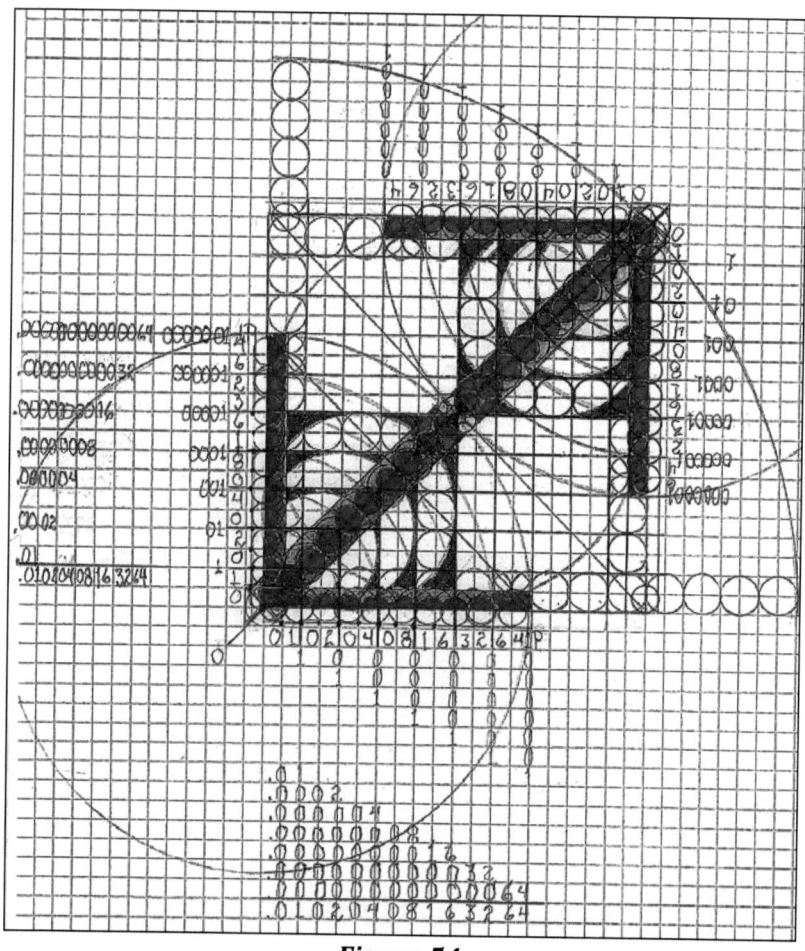

Figure 54

The Zero Armature for rotation of each $.5_{zi}$ x $.5_{ZI}$ with the $.70710678_{ZI}$ diagonal Hypotenuse is shown in Figure 55. The creation of the $.70710678_{ZI}$ in each quadrant is shown in Figure 56.

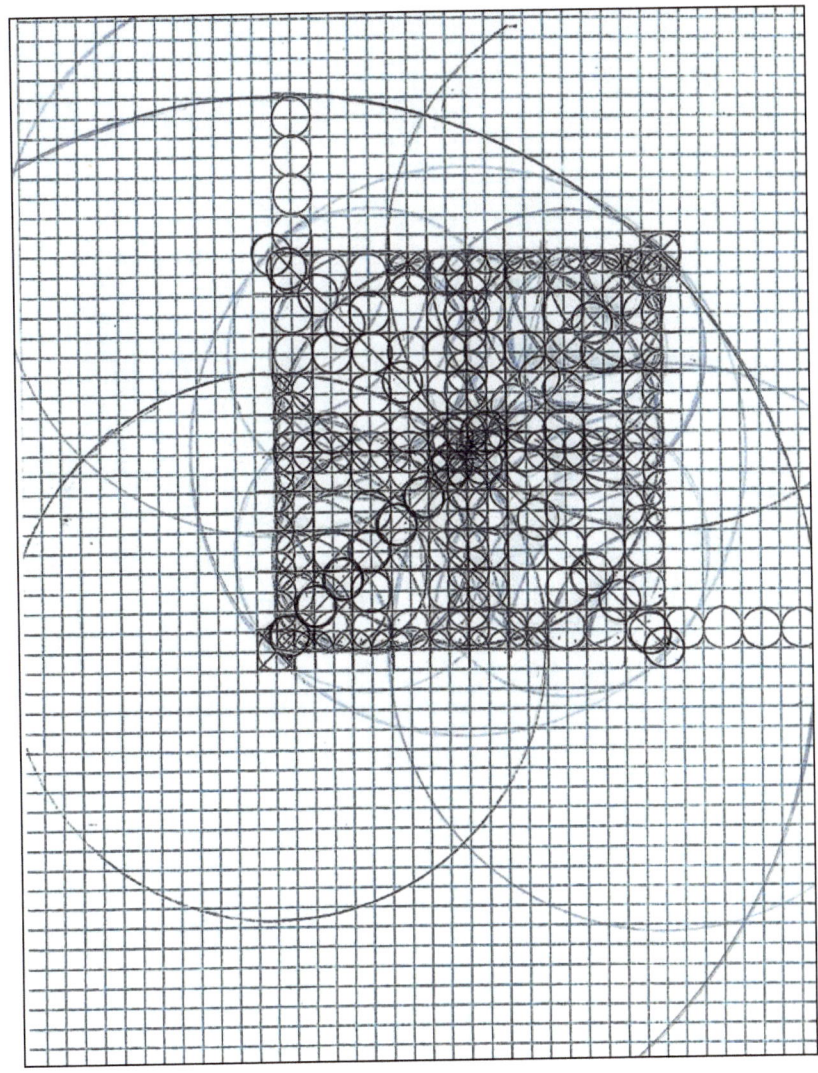

Figure 55

As the $.707106781_{ZI}$ Hypotenuse rotate from vertical and horizontal to the diagonal a $.5_{ZI}$ number is formed within each quadrant. The counts of N_Z, N_{ZI} and N_I are shown in Figure 56. The number formed does not rotate. Each quadrant square may rotate.

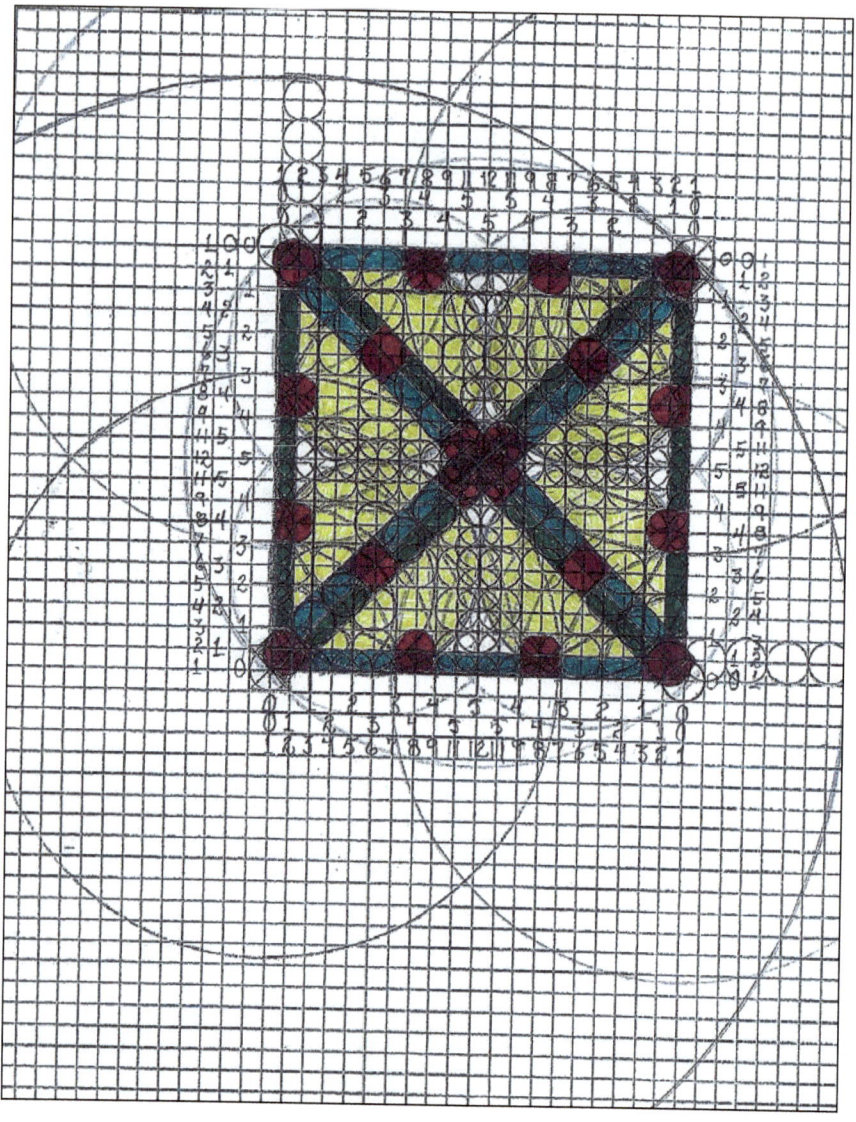

Figure 56

A Number is also formed within the 10 x 10 square as each .707106781 Hypotenuse is rotated from their vertical and horizontal positions. This creates a number within the 10 x 10 square that contains 2 Hypotenuse of 1.414213562_{ZI} as shown in Figure 57. The N_Z, N_{ZI}, and N_I counts are shown in Figure 57. This number does not rotate within the square, the entire 10 x 10 square may rotate.

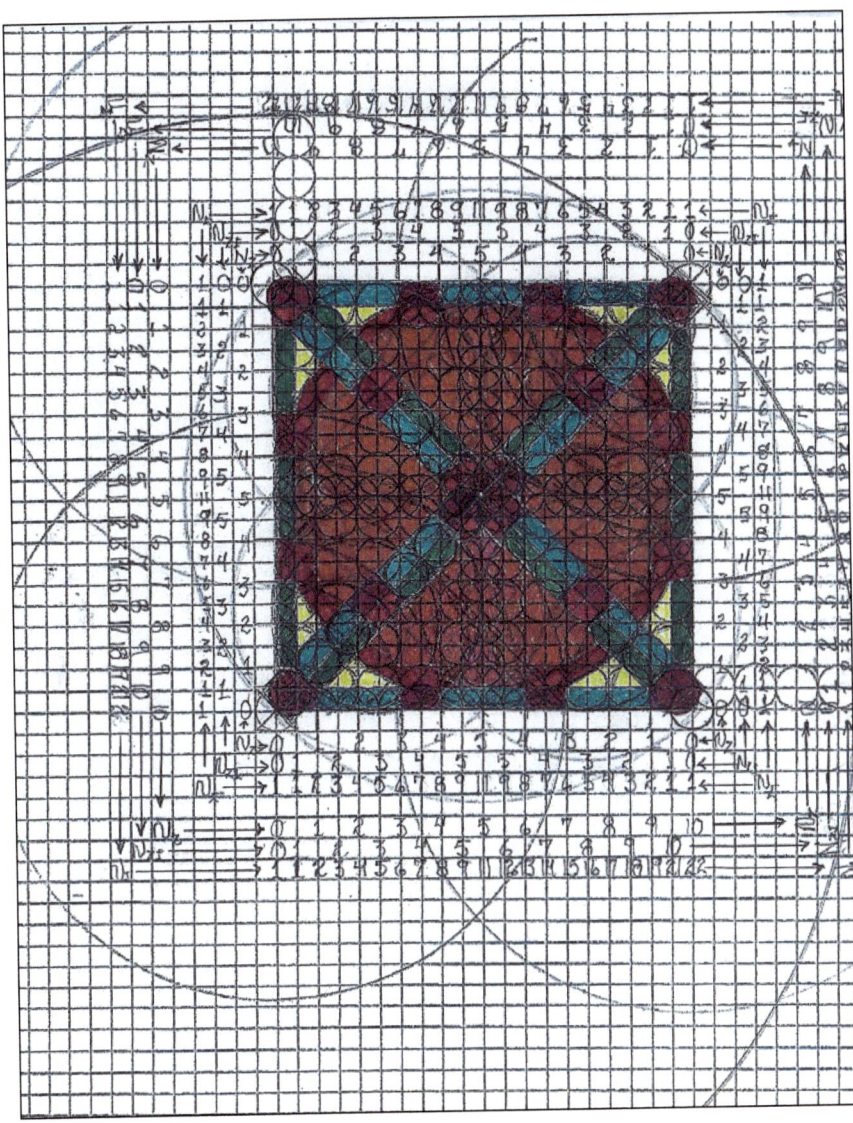

Figure 57

Each quadrant is now squared. This results in each quadrant being a $.5_{ZI}$ x $.5_{ZI}$ or $.25_{ZI}$. When each quadrant is added the result is 1.0_{ZI}. When all the .1 squares of a 10 x 10 square are added the result is 10.0_{ZI}. The entire 10 x 10 square may rotate (Figure 58).

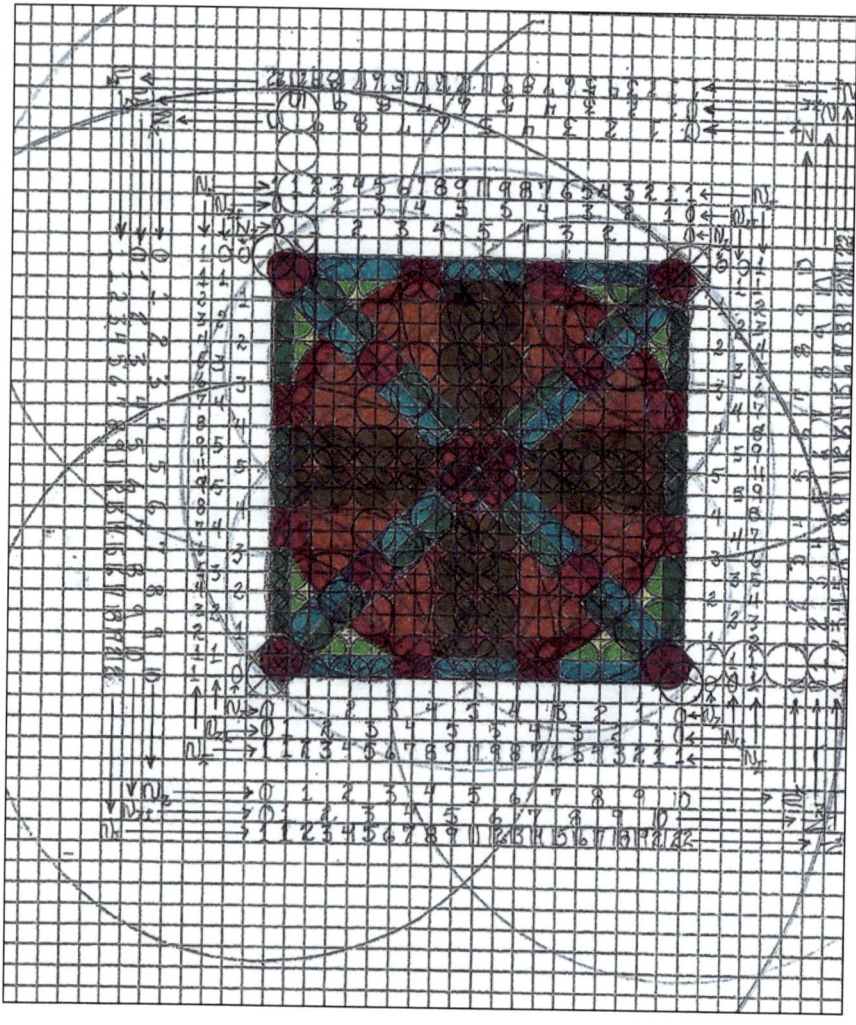

Figure 58

An Infinity-Based (N_{ZI}) number is created within each quadrant and the entire square. The numbers within each quadrant and within the entire square may now rotate as shown in Figure 59.

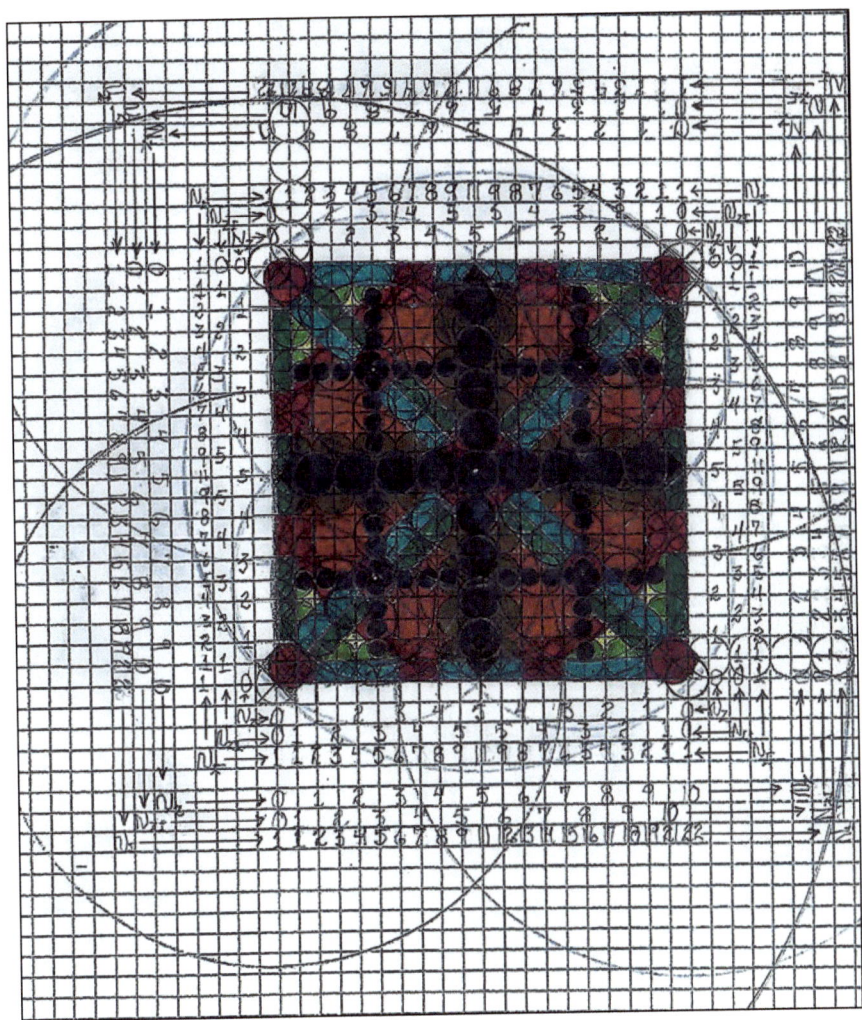

Figure 59

Chapter 6
Trigonometric Identities

All numbers are comprised of Ones. All Fractions are the result of a "Fractured" One. A fraction is added to the number or Zero The symbol used is . ,which is currently defined as a Decimal Point. The mathematical operations of Fractions and Whole numbers result in quantities that appear to be the same but are not. For example: .5 x .5 = .25 which is ½ of .5, while 5 x 5 = 25 which is 5 times 5. The failure to consider the difference has resulted in the number 1.414213562 being used as the square root of 2.

Figure 60 shows the rotation of the 14.14213562_{ZI} count. The rotation continues until it is encompassed by the Decimal 10_Z x 10_Z Counting Board. The lower-left corner is the axis of rotation, it is the center of the Prime One. The ZI count is along the diagonal to the upper-right corner. The number 14.14213562_{ZI} is multiplied by itself (14.14213562^2), using current computational devices results in a value of 199.99999999_{ZI}, when divided by 100_Z (10_Z x 10_Z square) the result is a value of 1.999999999_I or $1.414213562^2 = 2$.

$\sqrt{2} \neq 1.414213562$ and $1.414213562^2 \neq 2.0$

The mathematical calculations relative to the number of 1.414213562 are as follows:

$1^2.414213562^2 = 1 + .171572875 = 1.171572875$
$14^2.14213562^2 = 196 + .020202534 = 196.020202534$
$141^2.4213562^2 = 19881 + .177541047 = 19881.177541047$
$1414^2.213562^2 = 1999396 + .045608728 = 199396.045608728$
$14142^2.13562^2 = 199996164 + .018392784 = 199996164.01832784$
$141421^2.3562^2 = 1999989924 + .12687866 = 1999989924.12687866$
$1414213^2.562^2 = 1999998409 + .315844 = 1999998409.315844$

$14142135^2.62^2 = 1999999824 + .3844 = 1999999824.3844$

$14142135 6^2.2^2 = 1999999993 + .04 = 199999999.04$

$1414213562^2 = 1999999999 + 0 = 1999999999$

$1999999999/1000000000 = 1.999999999$

$\sqrt{1} + \sqrt{.999999999} = \sqrt{1} + \sqrt{1} = \sqrt{2}$

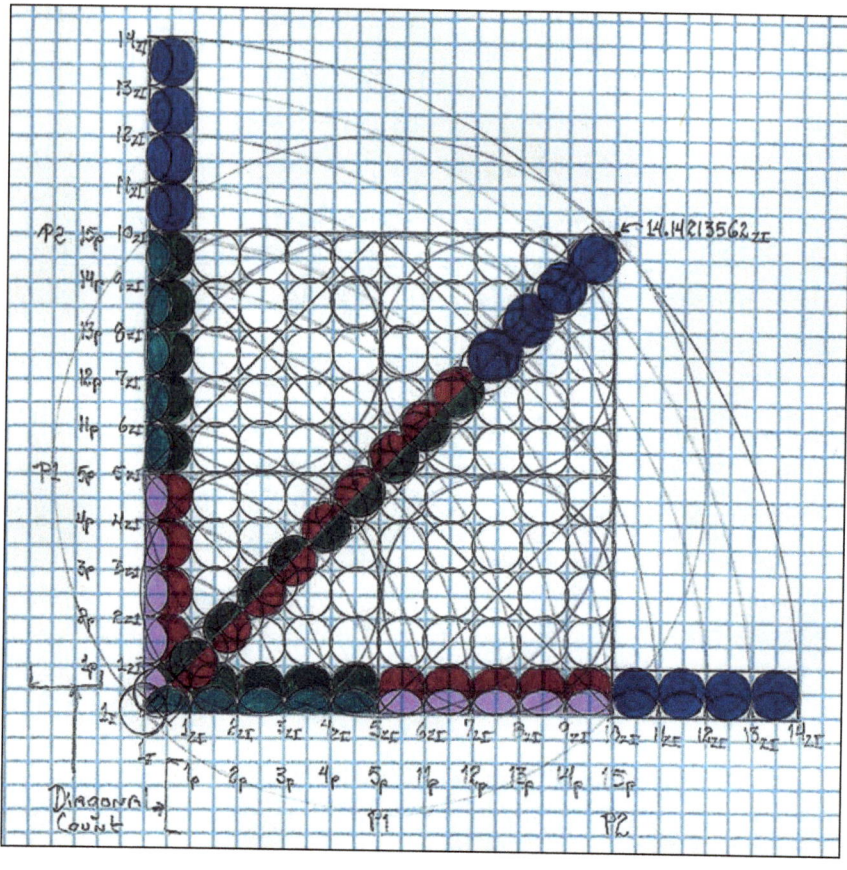

Figure 60

The count shown in Figure 60 is a positional count. The Vertical and Horizontal count of 14.14213562 exceeds the 10_Z x 10_Z square. Figure 61 shows the rotation of a quantifiable count. The horizontal and vertical count is $7_{ZI} + 7_{ZI}$ divided by 2. When each is rotated to the diagonal the diagonal count becomes 7.07106781_{ZI} +

$7.07106781_{ZI} = 14.14213562_{ZI}$. Shown within each diagonal number is 1_I, 3_I, 5_I, and 7_I, these are single-digit Prime numbers. The numbers, as shown in Figure 61, that are contained within the Decimal 10_Z x 10_Z square are diagonal 1_I, 3_I and 5_I.

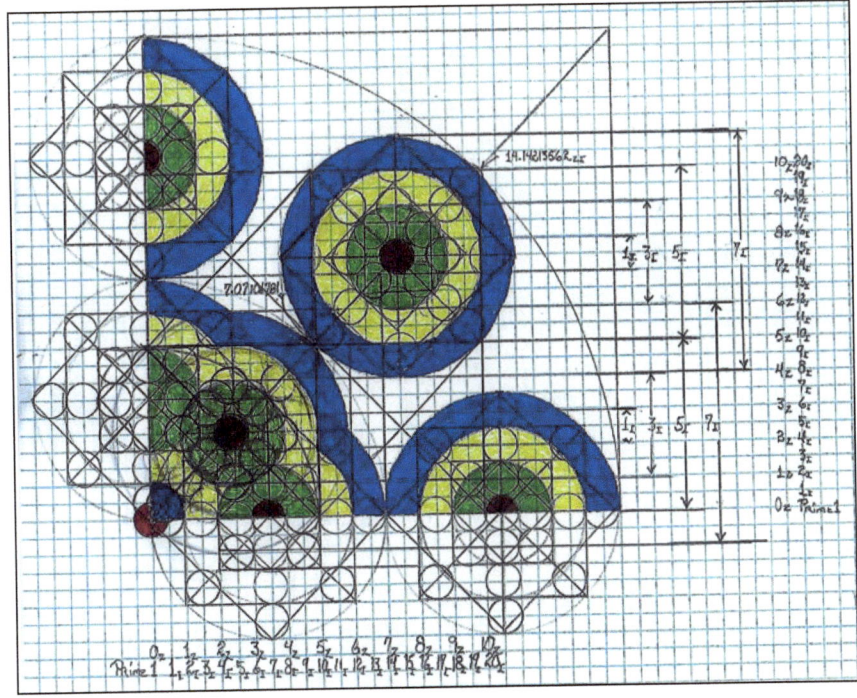

Figure 61

The counterclockwise rotation of ½ of the horizonal numbers and the clockwise rotation of ½ of the vertical numbers result in two whole numbers along the diagonal as shown in Figure 61. The two diagonal 5_I are contained within the 10_Z x 10_Z square. A 5_{ZI} diameter is multiplied either vertically or horizontally by itself, resulting in a 25_{ZI} square (5_{ZI} vertical diameter x 5 = 25_{ZI}, or 5_{ZI} horizontal diameter x 5 = 25_{ZI}.) Contained within each whole number 5_{ZI} is a diagonally squared 3_I, when counted along the vertical or horizontal diameter it becomes π. The counts are shown in Figure 62.

The Awakening of Numbers | 61

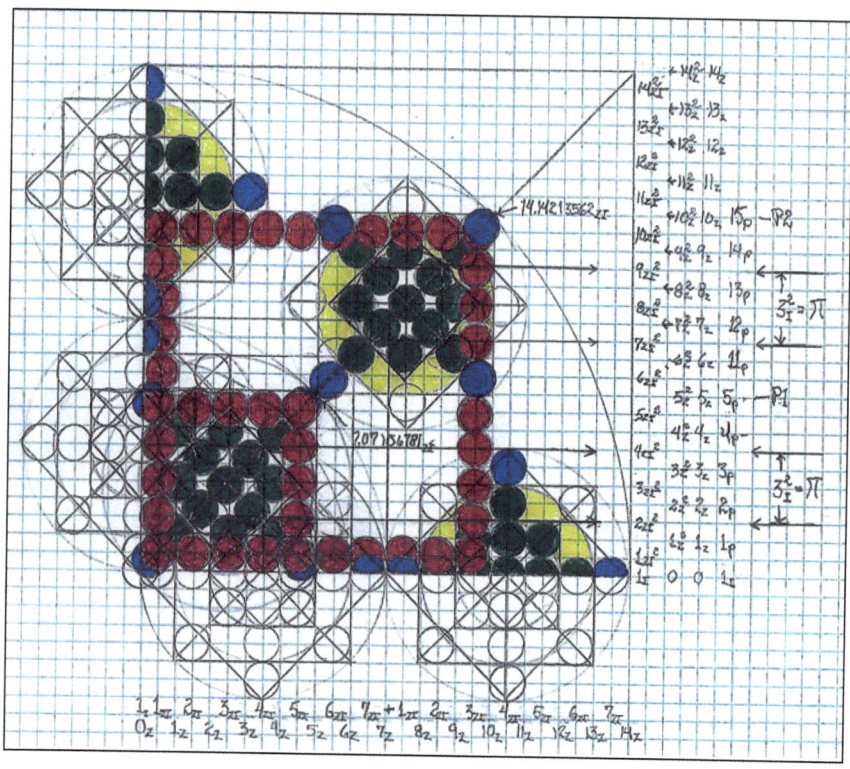

Figure 62

Shown in Figure 63 are the Armature components used for determining Trigonometric Identities of Sine, Cosine and Tangent. The Armature components are: Adjacent, Opposite, Hypotenuse, the positional measurements of the components result in Sine, Cosine and Tangent. The Sine, Cosine and Hypotenuse linear measurements are mathematically determined using the Pythagorean equation ($a^2 + b^2 = c^2$).

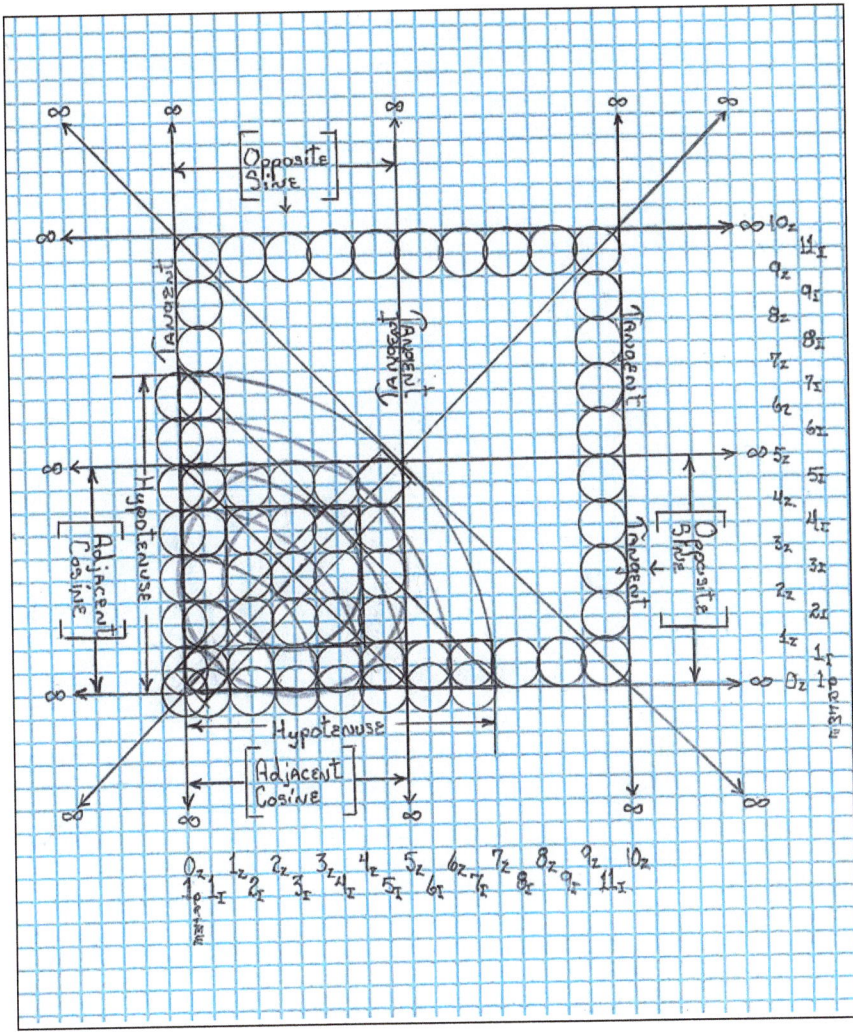

Figure 63

Figure 64 depicts the creation of Degrees. The initial N_Z count begins along the right vertical line of a 10_Z x 10_Z square, the count begins in the lower-right corner at 0_Z and proceeds upward in 5_Z increments. The count ends at the upper-right corner at a count of 50_Z. The count to 50_Z is diagonally divided in half, resulting in a count of 40_Z to 45_Z. This results in the upper-right corner being designated as

45. The initial Degree count is now 0 to 45 and is divided by 5 resulting in each 5_Z containing 9. The vertical and horizontal counts, as shown, depict the expansion of the Prime One to Infinity.

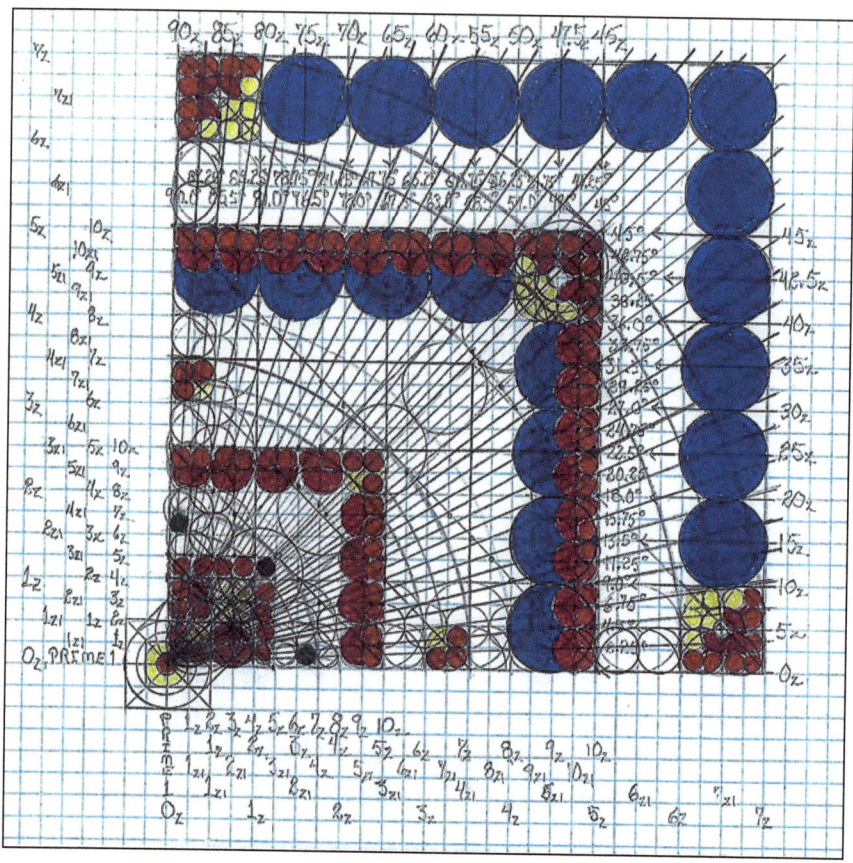

Figure 64

Figure 65 shows the counterclockwise rotation of the 7.07106781$_{PZI}$ Hypotenuse from 0 to 90.

Figure 65

Figure 66 shows the rotation of 7.07106781_{PZI} starting at 0_Z, 10_I and 7.07106781_P. The mathematical computations are:

Sine = Opposite (N_Z) ÷ Hypotenuse (N_P)
Cosine = Adjacent (N_I) ÷ Hypotenuse (N_P)
Tangent = Opposite (N_Z) ÷ Adjacent (N_I)

As the 7.07106781_P Hypotenuse is rotated counterclockwise, it cannot rotate clockwise because the Horizontal line is Zero (No numbers) meaning there can be no Numbers less than Zero. As the Hypotenuse rotates measurements are made along the vertical and horizontal lines and mathematically converted to Sine, Cosine and Tangent. All Degrees are measured within the Prime One, this results in all circles having 360°. Sine and Cosine are also measured within the Prime One and are always a fraction. The Tangent begins at the junction of the horizontal 10_Z Zero corner and proceeds vertically to infinity. Tangent measurements are made along these vertical lines.

As depicted in Figure 66 the beginning upward rotation to 45° results in a decrease in the Cosine count from 10_I to 7.07106781_I and an increase in the vertical Sine count of 0_Z to 7.07106781_Z. The count is divided by 10 resulting in the value being less than 1.

The count on the upper horizontal line, from 45° to 90° becomes the Sine count from 7.07106781_Z to 10_Z. The left vertical line becomes the Cosine Count from 7.07106781_I to 0_I. The count is divided by 10 resulting in the value being less than 1.

The Tangent count begins at 0_Z and continues vertically to 45° and results in a count of 10_Z. The count continues vertically from 45° to 90° from 10_Z to ∞. The count is divided by 10.

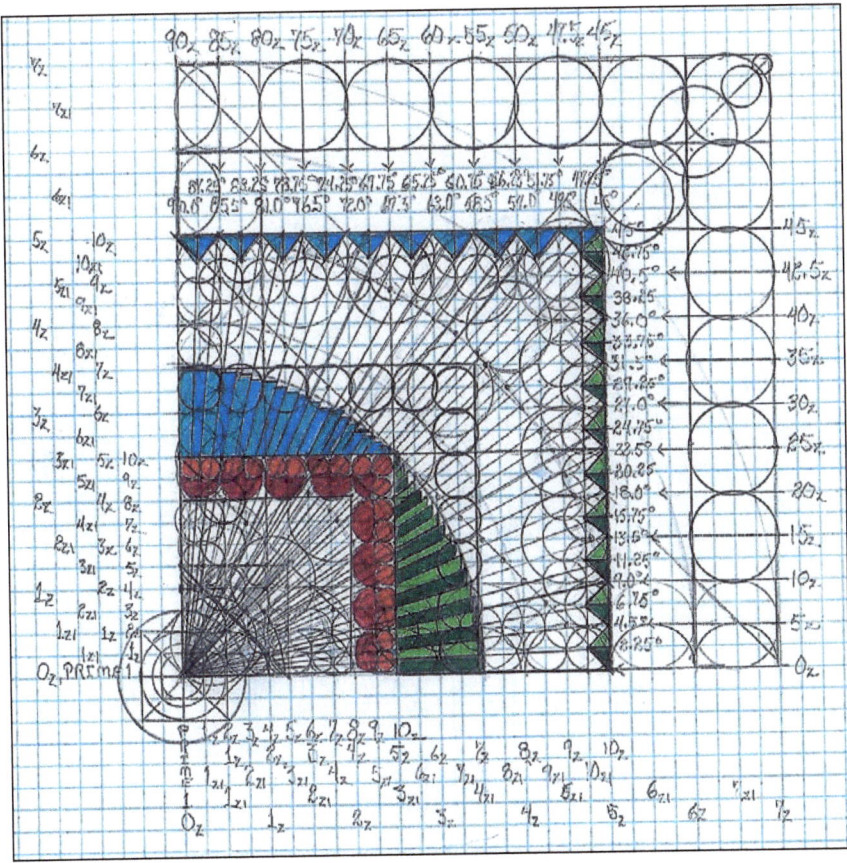

Figure 66

Figure 67 shows the Trigonometric Identity construct expansion from the Prime One and the corresponding counts. The expansion continues to Infinity.

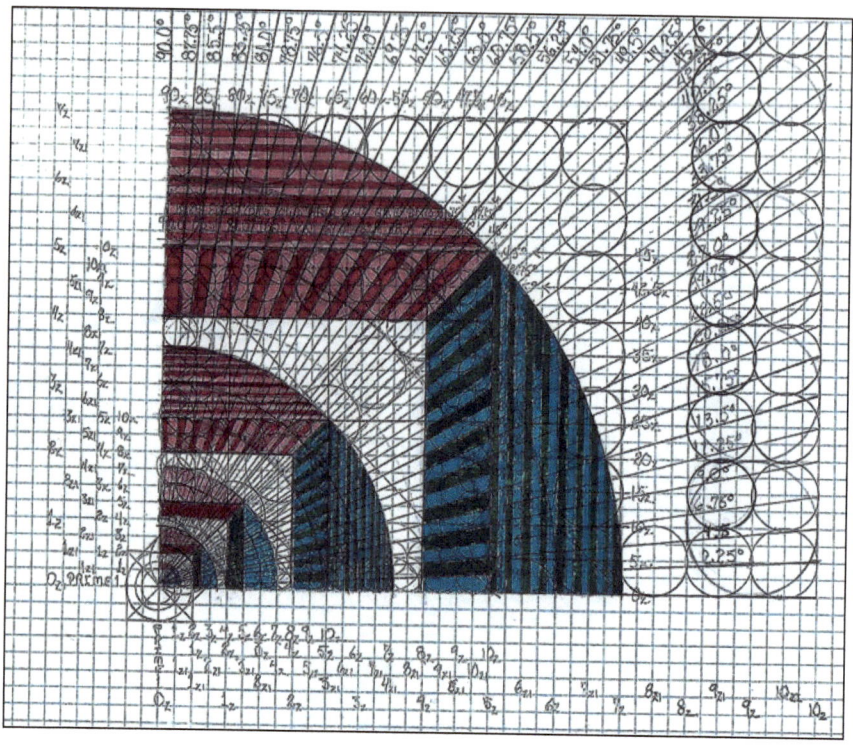

Figure 67

Table 1 contains the Pythagorean Equation results associated with the Trigonometric Identity constructs.

> (Note: The Pythagorean Equation is also known as the Pythagorean Theorem.)

The diagonal within the Pythagorean number is rotated forming 2 sides of a square.

The Sine count is increasing the Cosine count is decreasing.

$a^2 = (\text{Cosine} \bullet \text{Hypotenuse})^2$, $b^2 = (\text{Sine} \bullet \text{Hypotenuse})^2$, $c^2 = $ Hypotenuse^2, Hypotenuse = 7.07106781

$a^2 + b^2 = c^2$

Table 1:

Degree	Sine	Cosine	Hypotenuse	a²	+	b²	=	c²
00.0°	0_Z	1_I	7.0710678150_I		+	0_Z	=	50_{ZI}
02.5°	$.043619387_Z$	$.999048222_I$	7.07106781	49.90486743_I	+	$.095132548_Z$	=	50_{ZI}
05.0°	$.087155743_Z$	$.996194698_I$	7.07106781	49.6201938_I	+	$.379806174_Z$	=	50_{ZI}
07.5°	$.130526192_Z$	$.991444861_I$	7.07106781	49.14814563_I	+	$.851854339_Z$	=	50_{ZI}
10.0°	$.173648178_Z$	$.984807753_I$	7.07106781	48.49231549_I	+	1.507684485_Z	=	50_{ZI}
12.5°	$.216439614_Z$	$.976296007_I$	7.07106781	47.65769465_I	+	2.342305324_Z	=	50_{ZI}
15.0°	$.258819045_Z$	$.9655925826_I$	7.07106781	46.65063507_I	+	3.349364901_Z	=	50_{ZI}
17.5°	$.3007058_Z$	$.953716951_I$	7.07106781	45.47880111_I	+	4.521198905_Z	=	50_{ZI}
20.0°	$.342020143_Z$	$.939692621_I$	7.07106781	44.15111105_I	+	5.848888903_Z	=	50_{ZI}
22.5°	$.382683432_Z$	$.923879533_I$	7.07106781	42.67766951_I	+	7.322330453_Z	=	50_{ZI}
25.0°	$.422618262_Z$	$.906307787_I$	7.07106781	41.06969022_I	+	8.930309764_Z	=	50_{ZI}
27.5°	$.416748613_Z$	$.887010833_I$	7.07106781	39.33941089_I	+	10.66058907_Z	=	50_{ZI}
30.0°	$.5_Z$	$.866025404_I$	7.07106781	37.49999998_I	+	12.49999999_Z	=	50_{ZI}
32.5°	$.537299608_Z$	$.843391446_I$	7.07106781	35.56545652_I	+	14.43454343_Z	=	50_{ZI}
35.0°	$.573576436_Z$	$.819152044_I$	7.07106781	33.55050357_I	+	16.44949639_Z	=	50_{ZI}
37.5°	$.608761429_Z$	$.79335334_I$	7.07106781	31.47047611_I	+	18.53650781_Z	=	50_{ZI}
40.0°	$.64278761_Z$	$.766044443_I$	7.07106781	29.34120443_I	+	20.65879557_Z	=	50_{ZI}
42.5°	$.675590208_Z$	$.737277337_I$	7.07106781	27.17889355_I	+	22.82110645_Z	=	50_{ZI}
45.0°	$.707106781_Z$	$.707106781_I$	7.0710678125_I	25_I	+	25_Z	=	50_{ZI}

The Hypotenuse is now diagonal from the lower-left corner to the upper-right corner. The Hypotenuse continues to rotate in a counter-clockwise direction. The upper horizontal line is designated as Opposite, and the left vertical line is designated as Adjacent.

The Sine count continues to increase the Cosine count continues to decrease. The count continues as follows:

Degree	Sine	Cosine	Hypotenuse	a²	+	b²	=	c²
47.5°	.737277337$_Z$	.675590208$_I$	7.07106781	22.82110642$_I$	+	27.17889355$_Z$	=	50$_{ZI}$
50.0°	.766044443$_Z$	.64278761$_I$	7.07106781	20.65879557$_I$	+	29.34120443$_Z$	=	50$_{ZI}$
52.5°	.7933533$_Z$	.608761429$_I$	7.07106781	18.53650781$_I$	+	31.47047611$_Z$	=	50$_{ZI}$
55.0°	.819152044$_Z$	.573576436$_I$	7.07106781	16.53650781$_I$	+	33.55050357$_Z$	=	50$_{ZI}$
57.5°	.843391446$_Z$	.537299608$_I$	7.07106781	14.43459343$_I$	+	35.56545652$_Z$	=	50$_{ZI}$
60.0°	.866025404$_Z$	.5$_I$	7.07106781	12.49999999$_I$	+	37.49999998$_Z$	=	50$_{ZI}$
62.5°	.887010833$_Z$	.461748613$_I$	7.07106781	10.66058907$_I$	+	39.33941089$_Z$	=	50$_{ZI}$
65.0°	.906307787$_Z$	.422618262$_I$	7.07106781	8.930309764$_I$	+	41.06969022$_Z$	=	50$_{ZI}$
67.5°	.923879533$_Z$	.382683432$_I$	7.07106781	7.322330453$_I$	+	42.67766951$_Z$	=	50$_{ZI}$
70.0°	.939692621$_Z$	.342020143$_I$	7.07106781	5.848888908$_I$	+	44.15111105$_Z$	=	50$_{ZI}$
72.5°	.953716951$_Z$	.3007058$_I$	7.07106781	4.521198905$_I$	+	45.47880111$_Z$	=	50$_{ZI}$
75.0°	.965925826$_Z$	.258819045$_I$	7.07106781	3.349364901$_I$	+	46.65063507$_Z$	=	50$_{ZI}$
77.5°	.976296007$_Z$	.216439614$_I$	7.07106781	2.342305324$_I$	+	47.65769465$_Z$	=	50$_{ZI}$
80.0°	.984807753$_Z$	.173648178$_I$	7.07106781	1.507684485$_I$	+	48.49231549$_Z$	=	50$_{ZI}$
82.5°	.991444861$_Z$	.130526192$_I$	7.07106781	.851854559$_I$	+	49.14814563$_Z$	=	50$_{ZI}$
85.0°	.996194698$_Z$	.087155743$_I$	7.07106781	.379806174$_I$	+	49.6201938$_Z$	=	50$_{ZI}$
87.5°	.999048222$_Z$	.043619387$_I$	7.07106781	.095132548$_I$	+	49.90486743$_Z$	=	50$_{ZI}$
90.0°	1$_Z$	0	7.07106781	0	+	50$_Z$	=	50$_{ZI}$

The rotation within the square depicted in Figure 67 is complete, the entire square is now rotated. The rotation continues counterclockwise. Polarity is assigned to indicate the measurement within the respective quadrant. Reference Figure 68.

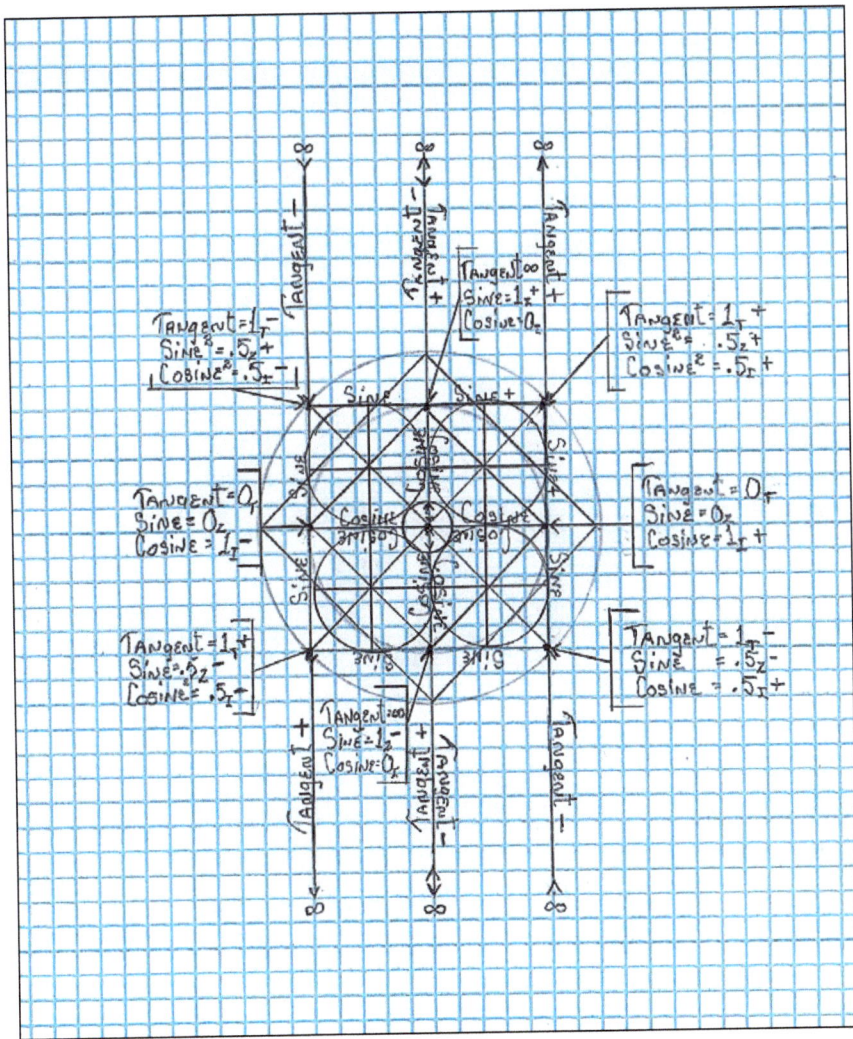

Figure 68

Upper-Right Quadrant:

The rotation of the 7.07106781_{ZI} Hypotenuse results in a vertical and horizontal square as show in Figure 68. The count is show in Figure 67 and Table 1.

The Sine starts at 0_Z and increases to 1_Z resulting in a positive (+) number.

The Cosine starts at 1_I and is a decreasing positive (+) number.

The Tangent line increases from 0_T to ∞ in an upward (+) direction.

Upper-Left Quadrant:

The entire initial square is rotated counterclockwise 90° resulting in a rectangle of 0° to 180° and is comprised of two 90° quarter-circle arcs. The configuration resembles a protractor that was developed when the Earth was believed to be flat.

The rotation of the 7.07106781_{ZI} Hypotenuse continues from 90° to 180°.

The Sine continues from 1_Z to 0_Z and is a decreasing positive (+) number.

The Cosine continues from 0_I to 1_I and is an increasing negative (-) number.

The Tangent line decreases from ∞ to 0_T in a downward (-) direction.

Lower-Left Quadrant:

The entire initial square is again rotated counterclockwise 90°.

The rotation of the 7.07106781_{ZI} Hypotenuse continues from 180° to 270°.

The Sine continues from 0_Z and is increasing to a negative 1_Z (-).

The Cosine continues from negative 1_I (-) to 0_I and is a decreasing negative (-) number.

The Tangent line increases from 0_T to ∞ in a downward direction (-).

Lower-Right Quadrant:

The entire initial square is again rotated counterclockwise 90°.

The rotation of the 7.07106781_{ZI} Hypotenuse continues from 270° to 360°.

The Sine continues from negative 1_Z (-) to 0_Z and is a decreasing negative (-) number.

The Cosine continues from 0_I and is increasing (+) to a positive (+) 1_I.

The Tangent line decreases from ∞ to 0_T in an upward (-) direction.

The entire square has rotated 360°, the Hypotenuse initial starting point was 2.5° and now continues from 360° to 2.5° completing the circle. This is the initial formation of this circle all subsequent circles are 360°. When the initial Decimal circle is formed the Degree, Sine, Cosine and Tangent numbers remain as expressed here, the starting point for the initial formation of a Decimal circle, which is 5_Z. The resulting 365° was converted to 365 days.

Chapter 7
Number Uses

Calendars

Figure 69 is a construct that shows numbers, within the Decimal system that are used, by Humans of European ancestry, daily. There are many things that can be derived from studying the One, only a few will be elaborated upon.

Figure 69

Shown in Figure 69 is the 12-month calendar, used by Humans of European ancestry, and a comparison with the 10-month calendar. The 10-month calendar is a function of the Decimal One and is annotated as such. The 12-month calendar is a function of the Decimal Number 3. It is observed that there are two additional months, July and August. It is believed that this is the origin of the Julian Calendar. Julius Caesar gets credit for the discovery of the 12-month calendar. His adopted son Augustus Caesar, who also considered himself a God, incorporates/changes the names of the two additional months to July and August, and places them immediately after the previous Gods. The name of the calendar is then designated as the Julian calendar, in honor of his father. Later Pope Gregory discovers the Earth's rotation has changed since the development of the calendar, by the Romans. Pope Gregory invokes the findings and takes credit for them. The Leap year now comes into existence. The Julian calendar changes to the Gregorian calendar. The adjustment of the slowing of the Earth's rotation continues to date. Figure 70 shows the Calendar with the 24-hour clock incorporated, the rotation is clockwise indicating that time is constantly being divided/subtracted/used.

A set of Dice contains 2 Di that are cubes and Numbers 1 to 6 are contained on each Di. With the One being opposite of a Six. These are also part of the Decimal numbering system. There are also 52 cards in a deck and 52 weeks in a year.

The symbols and correlations are many in Human civilization and they vary depending on the culture and language. There are many symbols that are a result of the Decimal system constructs, many have religious or mystical connotations.

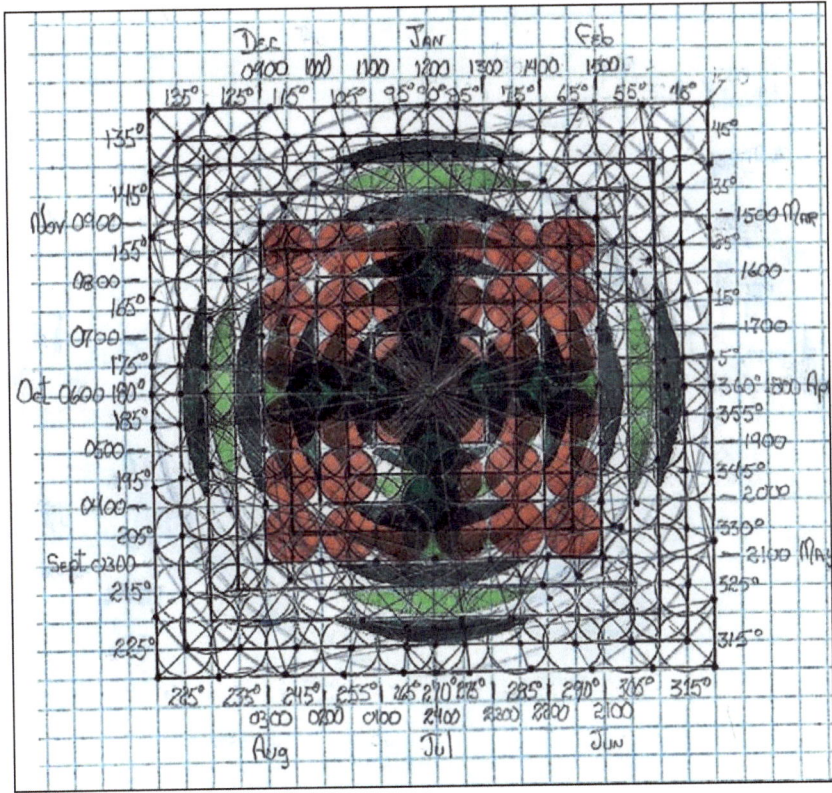

Figure 70

Pi:

Pi is referred to as a Constant. Pi as shown on a calculator, is 3.14592654.

The Pi equation is stated as π = Circumference/Diameter, and Pi is stated as a constant number. This is not consistent with Algebraic principles and methodology. When the equation is stated such that we solve for Pi the equation becomes Circumference/Diameter = π.

Solve for Pi using the 360-degree number. 360° = πD. At this point it must be determined what diameter is being used, the circumference is always measured in degrees. Using a Diameter of 180 and dividing 360° by 180° results in a π of 2 which is 2 diameters. When the diagonal Diameters are considered π becomes 4.

The Awakening of Numbers | 77

Diameter/Circumference (Geometry):

Diameters are a result of a number being squared. If a circle is assigned a diameter it means the circle is squared. It may be squared internally or externally, and it may be horizontal/vertical or diagonally. It is superimposed on a Counting Table.

We will begin Geometry with the construct shown in Figure 71. The vertical line will be rotated to the diagonal and the horizontal line to the diagonal. The vertical and horizontal circles have a diameter of 10, it is squared, this is a radius of the circle that will be formed as shown in Figure 71. The squared radius is a Counting table and it is used as originally developed, the count is 5 up to the peak and down 5 on the horizontal and is repeated on the vertical, as shown in Figure 72.

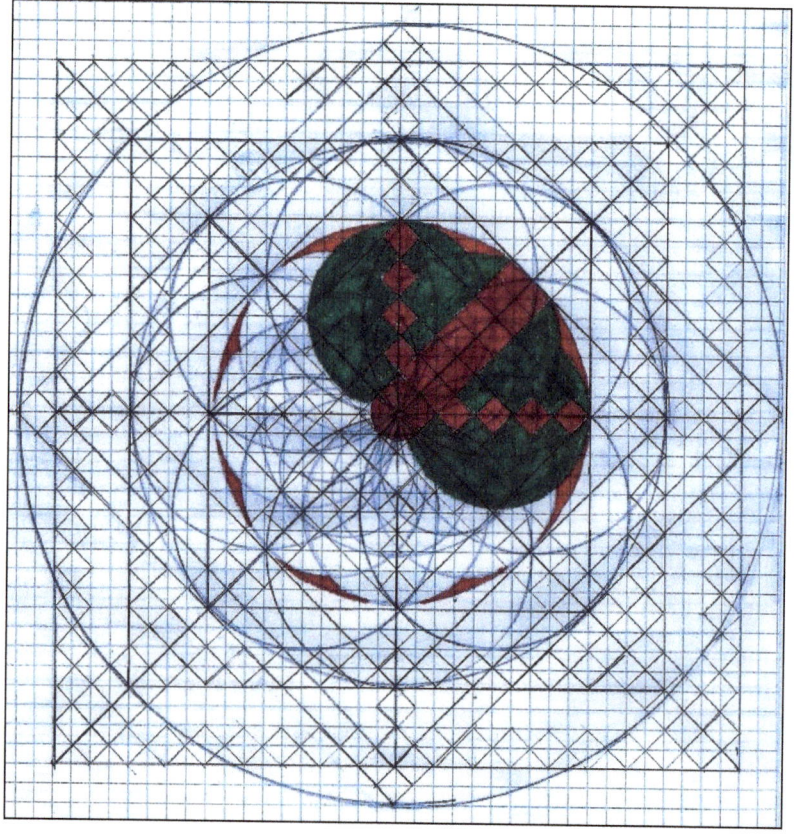

Figure 71

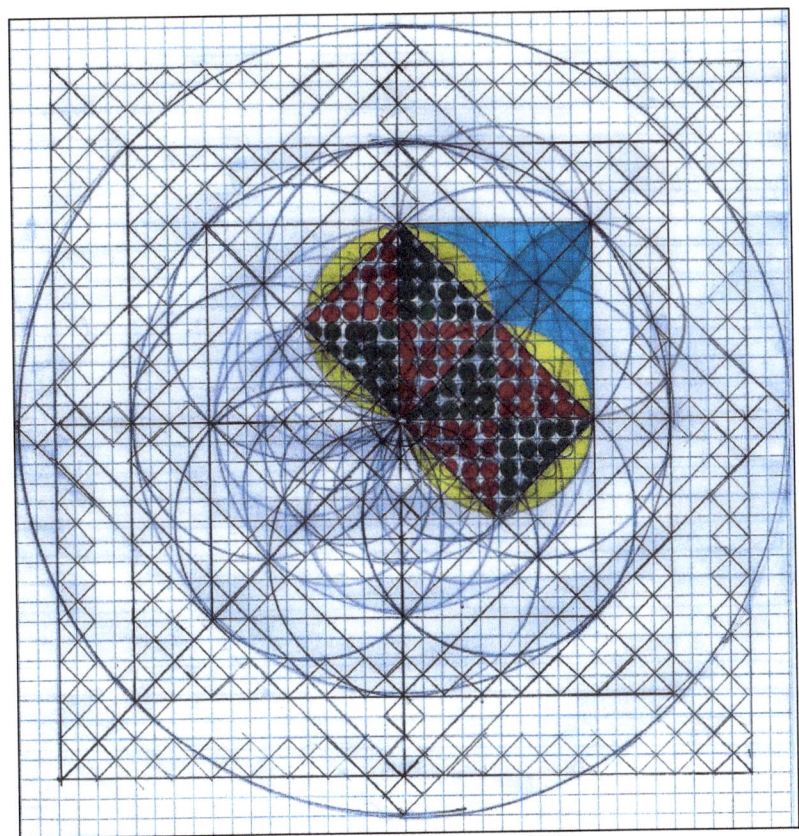

Figure 72

The first equation to be used to determine the Circumference is A+B = C, while holding C constant at a 10. This is an internally squared circle. The Radius (Hypotenuse) will proceed up one side of the triangle within the square, one step at a time and proceed down the adjacent triangle within the square, one step at a time. The Radius (Hypotenuse) will remain at a decimal 10_I from the axis point. As shown in Figure 73 the number that is being rotated is a 10_{ZI} number. The rotation continues through 4 quadrants.

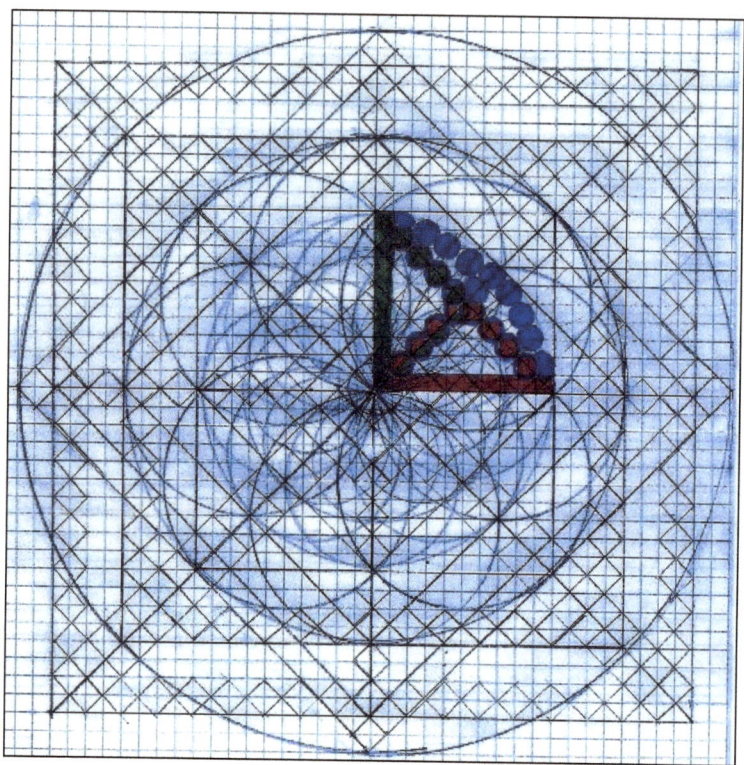

Figure 73

The count on the perimeter is 10/10 = 1, 10/9 = 1.111, 10/8 = 1.25, 10/7 = 1.428, 10/6 = 1.666, 10/5 = 2, this is up the opposite side of the triangle, 10/5 = 2, 10/6 = 1.666, 10/7 = 1.428, 10/8 = 1.25, 10/9 = 1.111, 10/10 = 1, this is down opposite side of the adjoining triangle.

360-Degree Geometry:

The Number 36 (360 degrees) will be used for explanation of the Number Equations associated with Geometry/Trigonometry. This concept is applicable by convention only; the equations are applicable to all numbers.

The diameter of a 360-degree circle is 180 degrees and the radii are 90 degrees. The Equation for Circumference is currently stated as: Circumference = πD. It is shown in Figure 74, it is 3^2 within a circle and is measured corner to corner.

Figure 74

Circular Area:

The equation for Circular Area is , when π is equal to a Decimal 4, the equation becomes A = 4 x radius². Figure 75 shows a radius of the entire infinity-based circle squared. Any one radius of the 8 contained within the circle may be used. The count is $r^2 + r^2 + r^2 + r^2$ or $4r^2$. Area is the squared Radius that moves in a circular manner. The square also contains a circle.

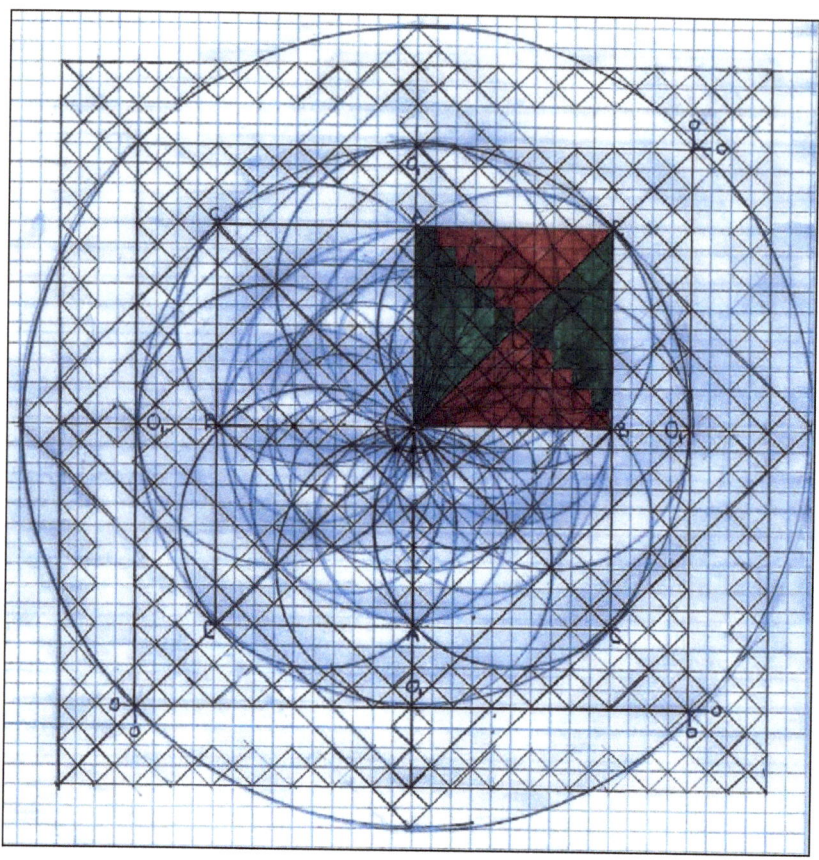

Figure 75

$a_{ZI} + b_{ZI} = c_{ZI}$

Geometry and Trigonometry are inseparable, meaning you cannot do one without doing the other. The analysis usually is done with a square within a circle and measurements are made within the square. a + b = c is the measurement of a circle within square and it is measured on the square.

a = Horizontal = adjacent. b = Vertical = opposite. c = Radius = hypotenuse.

The Number contains an internal square that is rotated. Beginning at the center of the circle and counting outward on a radius, the number that is rotated is 9_{ZI}. This count permits rotation of the square within the circle. The square within the circle is rotated and measured on the external square's perimeter that encloses the circle. Reference Figure 76.

The outer square is 10_Z x 10_Z, the resulting measurements are divided by 10. The results are expressed as fractions of the Prime One. Reference Table 2

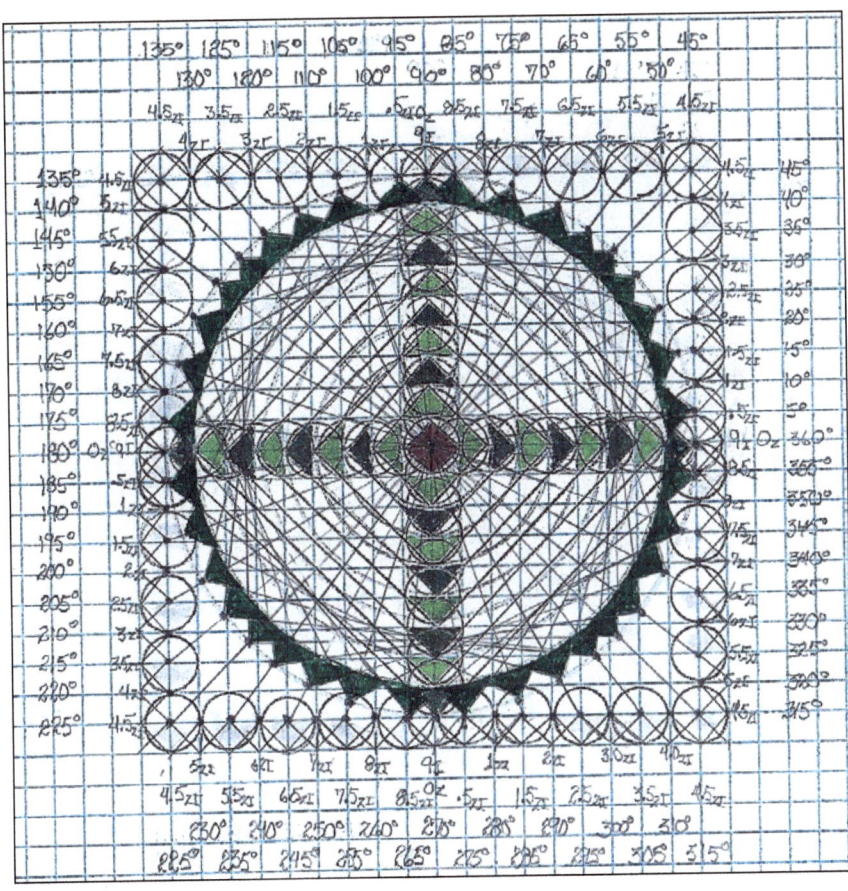

Figure 76

Table 2:
Quadrant 1

Degree	Adjacent (a)	+	Opposite (b)	=	Hypotenuse (c)	Tangent, b/a	Sine b/c	Cosine a/c
360	9$_{ZI}$	+	0$_{ZI}$	=	9$_{ZI}$	0.000000000	0.000000000	1.000000000
5	8.5$_{ZI}$	+	.5$_{ZI}$	=	9$_{ZI}$	.0588235290	.0555555556	.9444444444
10	8.0$_{ZI}$	+	1.0$_{ZI}$	=	9$_{ZI}$	.1250000000	.1111111111	.8888888889
15	7.5$_{ZI}$	+	1.5$_{ZI}$	=	9$_{ZI}$	.2000000000	.1666666666	.8333333333
20	7.0$_{ZI}$	+	2.0$_{ZI}$	=	9$_{ZI}$	.2857142860	.2222222222	.7777777777
25	6.5$_{ZI}$	+	2.5$_{ZI}$	=	9$_{ZI}$	.3846153850	.2777777777	.7222222222
30	6.0$_{ZI}$	+	3.0$_{ZI}$	=	9$_{ZI}$	.5000000000	.3333333333	.6666666666
35	5.5$_{ZI}$	+	3.5$_{ZI}$	=	9$_{ZI}$	.6363636363	.3888888888	.6111111111
40	5.0$_{ZI}$	+	4.0$_{ZI}$	=	9$_{ZI}$	.8000000000	.4444444444	.5555555556
45	4.5$_{ZI}$	+	4.5$_{ZI}$	=	9$_{ZI}$	1.000000000	.5000000000	.5000000000
50	4.0$_{ZI}$	+	5.0$_{ZI}$	=	9$_{ZI}$	1.250000000	.5555555556	.4444444444
55	3.5$_{ZI}$	+	5.5$_{ZI}$	=	9$_{ZI}$	1.571428571	.6111111111	.3888888888
60	3.0$_{ZI}$	+	6.0$_{ZI}$	=	9$_{ZI}$	2.000000000	.6666666666	.3333333333
65	2.5$_{ZI}$	+	6.5$_{ZI}$	=	9$_{ZI}$	2.600000000	.7222222222	.2777777777
70	2.0$_{ZI}$	+	7.0$_{ZI}$	=	9$_{ZI}$	3.500000000	.7777777777	.2222222222
75	1.5$_{ZI}$	+	7.5$_{ZI}$	=	9$_{ZI}$	5.000000000	.8333333333	.1666666666
80	1.0$_{ZI}$	+	8.0$_{ZI}$	=	9$_{ZI}$	8.000000000	.8888888888	.1111111111
85	.5$_{ZI}$	+	8.5$_{ZI}$	=	9$_{ZI}$	17.00000000	.9444444444	.0555555556
90	0$_{ZI}$	+	9.0$_{ZI}$	=	9$_{ZI}$	∞	1.000000000	0

Quadrant 2

Degrees	Adjacent (a)	+	Opposite (b)	=	Hypotenuse (c)	Tangent	Sine	Cosine
95	0.5$_{ZI}$	+	8.5$_{ZI}$	=	9$_{ZI}$	17.00000000	.9444444444	.0555555556
100	1.0$_{ZI}$	+	8.0$_{ZI}$	=	9$_{ZI}$	8.000000000	.8888888888	.1111111111
105	1.5$_{ZI}$	+	7.5$_{ZI}$	=	9$_{ZI}$	5.000000000	.8333333333	.1666666666
110	2.0$_{ZI}$	+	7.0$_{ZI}$	=	9$_{ZI}$	3.500000000	.7777777777	.2222222222
115	2.5$_{ZI}$	+	6.5$_{ZI}$	=	9$_{ZI}$	2.600000000	.7222222222	.2777777777
120	3.0$_{ZI}$	+	6.0$_{ZI}$	=	9$_{ZI}$	2.000000000	.6666666666	.3333333333
125	3.5$_{ZI}$	+	5.5$_{ZI}$	=	9$_{ZI}$	1.571428571	.6111111111	.3888888888
130	4.0$_{ZI}$	+	5.0$_{ZI}$	=	9$_{ZI}$	1.250000000	.5555555556	.4444444444

Degrees	Adjacent (a)	+	Opposite (b)	=	Hypotenuse (c)	Tangent	Sine	Cosine
135	4.5_{ZI}	+	4.5_{ZI}	=	9_{ZI}	1.000000000	.5000000000	.5000000000
140	5.0_{ZI}	+	4.0_{ZI}	=	9_{ZI}	.8000000000	.4444444444	.5555555556
145	5.5_{ZI}	+	3.5_{ZI}	=	9_{ZI}	6363636363	.3888888888	.6111111111
150	6.0_{ZI}	+	3.0_{ZI}	=	9_{ZI}	.5000000000	.3333333333	.6666666666
155	6.5_{ZI}	+	2.5_{ZI}	=	9_{ZI}	.3846153850	.2777777777	.7222222222
160	7.0_{ZI}	+	2.0_{ZI}	=	9_{ZI}	.2857142860	.2222222222	.7777777777
165	7.5_{ZI}	+	1.5_{ZI}	=	9_{ZI}	.2000000000	.1666666666	.8333333333
170	8.0_{ZI}	+	1.0_{ZI}	=	9_{ZI}	.1250000000	.1111111111	.8888888888
175	8.5_{ZI}	+	0.5_{ZI}	=	9_{ZI}	.0588235290	.0555555556	.9444444444
180	9.0_{ZI}	+	0_{ZI}	=	9_{ZI}	0	0	1.000000000

Quadrant 3

Degrees	Adjacent (a)	+	Opposite (b)	=	Hypotenuse (c)	Tangent	Sine	Cosine
185	8.5_{ZI}	+	$.5_{ZI}$	=	9_{ZI}	.0588235290	.0555555556	.9444444444
190	8.0_{ZI}	+	1.0_{ZI}	=	9_{ZI}	.1250000000	.1111111111	.8888888889
195	7.5_{ZI}	+	1.5_{ZI}	=	9_{ZI}	.2000000000	.1666666666	.8333333333
200	7.0_{ZI}	+	2.0_{ZI}	=	9_{ZI}	.2857142860	.2222222222	.7777777777
205	6.5_{ZI}	+	2.5_{ZI}	=	9_{ZI}	.3846153850	.2777777777	.7222222222
210	6.0_{ZI}	+	3.0_{ZI}	=	9_{ZI}	.5000000000	.3333333333	.6666666666
215	5.5_{ZI}	+	3.5_{ZI}	=	9_{ZI}	.6363636363	.3888888888	.6111111111
220	5.0_{ZI}	+	4.0_{ZI}	=	9_{ZI}	.8000000000	.4444444444	.5555555556
225	4.5_{ZI}	+	4.5_{ZI}	=	9_{ZI}	1.000000000	.5000000000	.5000000000
230	4.0_{ZI}	+	5.0_{ZI}	=	9_{ZI}	1.250000000	.5555555556	.4444444444
235	3.5_{ZI}	+	5.5_{ZI}	=	9_{ZI}	1.571428571	.6111111111	.3888888888
240	3.0_{ZI}	+	6.0_{ZI}	=	9_{ZI}	2.000000000	.6666666666	.3333333333
245	2.5_{ZI}	+	6.5_{ZI}	=	9_{ZI}	2.600000000	.7222222222	.2777777777
250	2.0_{ZI}	+	7.0_{ZI}	=	9_{ZI}	3.500000000	.7777777777	.2222222222
255	1.5_{ZI}	+	7.5_{ZI}	=	9_{ZI}	5.000000000	.8333333333	.1666666666
260	1.0_{ZI}	+	8.0_{ZI}	=	9_{ZI}	8.000000000	.8888888888	.1111111111
265	$.5_{ZI}$	+	8.5_{ZI}	=	9_{ZI}	17.00000000	.9444444444	.0555555556
270	0	+	9.0_{ZI}	=	9_{ZI}	∞	1.000000000	0

The Awakening of Numbers

Quadrant 4

Degrees	Adjacent (a)	+	Opposite (b)	=	Hypotenuse (c)	Tangent	Sine	Cosine
275	0.5$_{ZI}$	+	8.5$_{ZI}$	=	9$_{ZI}$	17.00000000	.9444444444	.0555555556
280	1.0$_{ZI}$	+	8.0$_{ZI}$	=	9$_{ZI}$	8.000000000	.8888888888	.1111111111
285	1.5$_{ZI}$	+	7.5$_{ZI}$	=	9$_{ZI}$	5.000000000	.8333333333	.1666666666
290	2.0$_{ZI}$	+	7.0$_{ZI}$	=	9$_{ZI}$	3.500000000	.7777777777	.2222222222
295	2.5$_{ZI}$	+	6.5$_{ZI}$	=	9$_{ZI}$	2.600000000	.7222222222	.2777777777
300	3.0$_{ZI}$	+	6.0$_{ZI}$	=	9$_{ZI}$	2.000000000	.6666666666	.3333333333
305	3.5$_{ZI}$	+	5.5$_{ZI}$	=	9$_{ZI}$	1.571428571	.6111111111	.3888888888
310	4.0$_{ZI}$	+	5.0$_{ZI}$	=	9$_{ZI}$	1.250000000	.5555555556	.4444444444
315	4.5$_{ZI}$	+	4.5$_{ZI}$	=	9$_{ZI}$	1.000000000	.5000000000	.5000000000
320	5.0$_{ZI}$	+	4.0$_{ZI}$	=	9$_{ZI}$	.8000000000	.4444444444	.5555555556
325	5.5$_{ZI}$	+	3.5$_{ZI}$	=	9$_{ZI}$	.6363636363	.3888888888	.6111111111
330	6.0$_{ZI}$	+	3.0$_{ZI}$	=	9$_{ZI}$	.5000000000	.3333333333	.6666666666
335	6.5$_{ZI}$	+	2.5$_{ZI}$	=	9$_{ZI}$	.3846153850	.2777777777	.7222222222
340	7.0$_{ZI}$	+	2.0$_{ZI}$	=	9$_{ZI}$	.2857142860	.2222222222	.7777777777
345	7.5$_{ZI}$	+	1.5$_{ZI}$	=	9$_{ZI}$	.2000000000	.1666666666	.8333333333
350	8.0$_{ZI}$	+	1.0$_{ZI}$	=	9$_{ZI}$	.1250000000	.1111111111	.8888888888
355	8.5$_{ZI}$	+	0.5$_{ZI}$	=	9$_{ZI}$	.0588235290	.0555555556	.9444444444
360	9.0$_{ZI}$	+	0$_{ZI}$	=	9$_{ZI}$	0	0	1.000000000

Decimal Zero-Based (10Z) Trigonometric Geometry (360)

Figure 77 is the beginning construct of Zero-based 360. As observed, it is also the initial construct of the Decimal numbering system. The green triangle is counted 5 up on the diagonal and 5 across on the horizontal and 5 up on the vertical. The red triangle is counted 5 down on the diagonal and 5 down on the vertical and 5 across on the horizontal. The numbers 1, 5 and 10 are the tracked numbers.

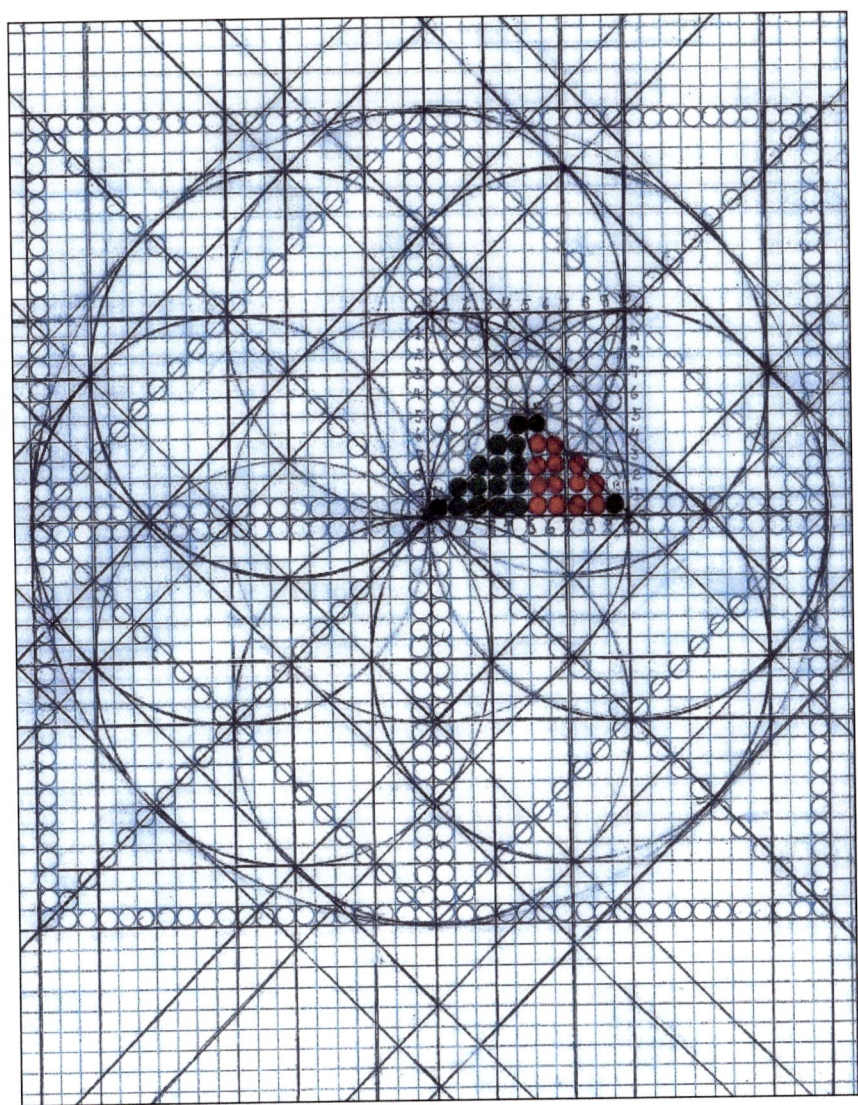

Figure 77

Figure 78 shows the first move of the red triangle up to the next quadrant.

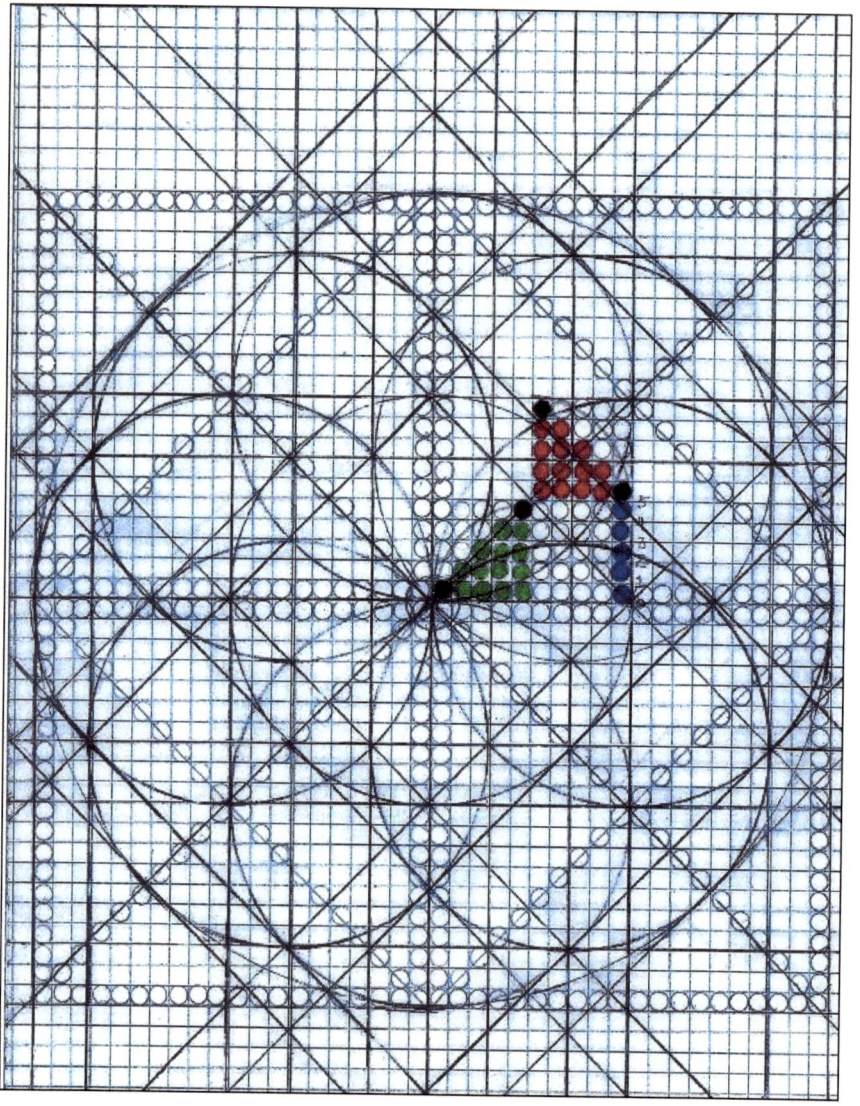

Figure 78

Figure 79 is the second move of red triangle; the pattern of movement is indicated by the blue circles.

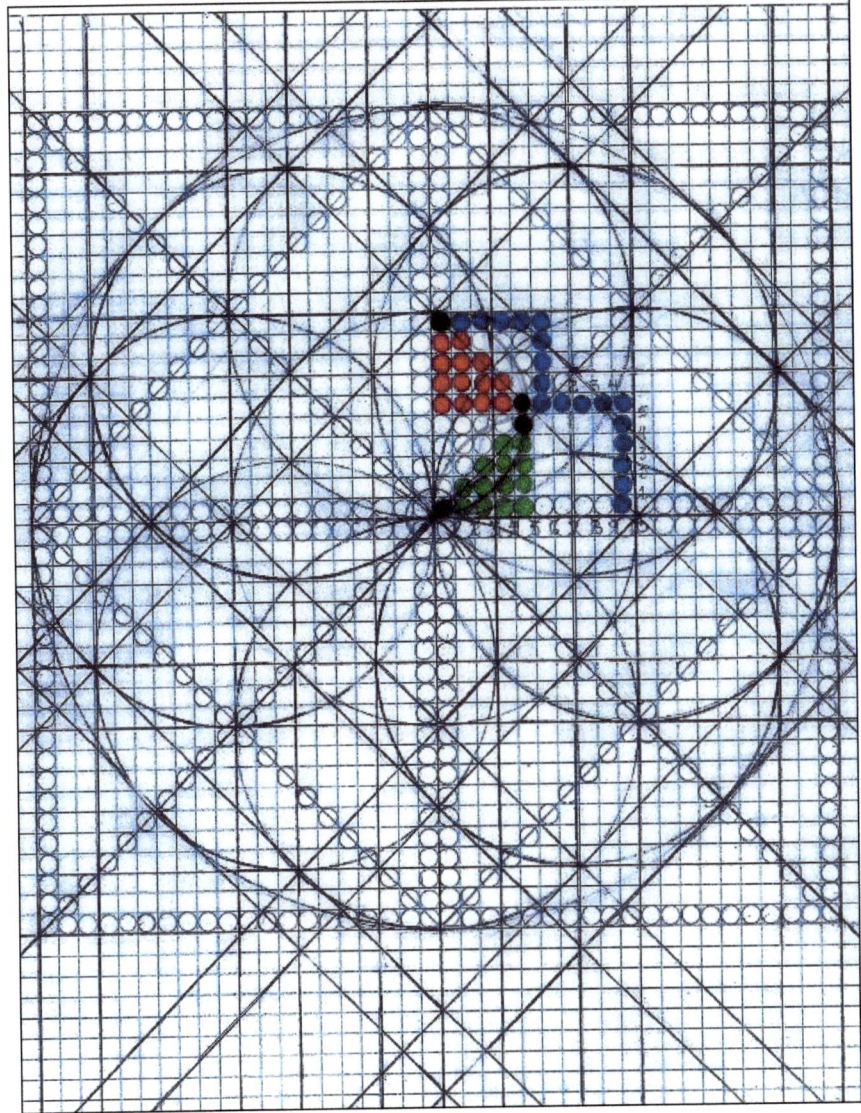

Figure 79

The green triangle now moves horizontally one quadrant and the red triangle moves down one quadrant as shown in Figure 80.

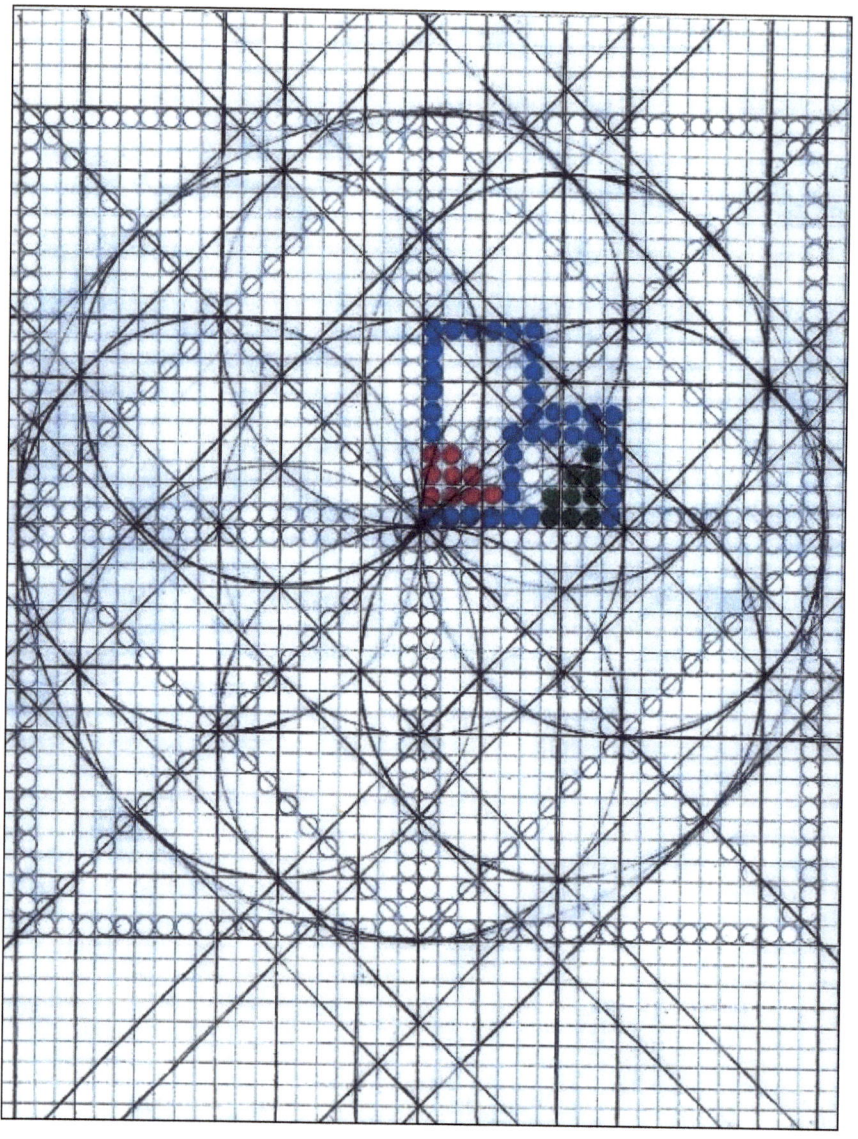

Figure 80

The green triangle now moves up one quadrant. Figure 81.

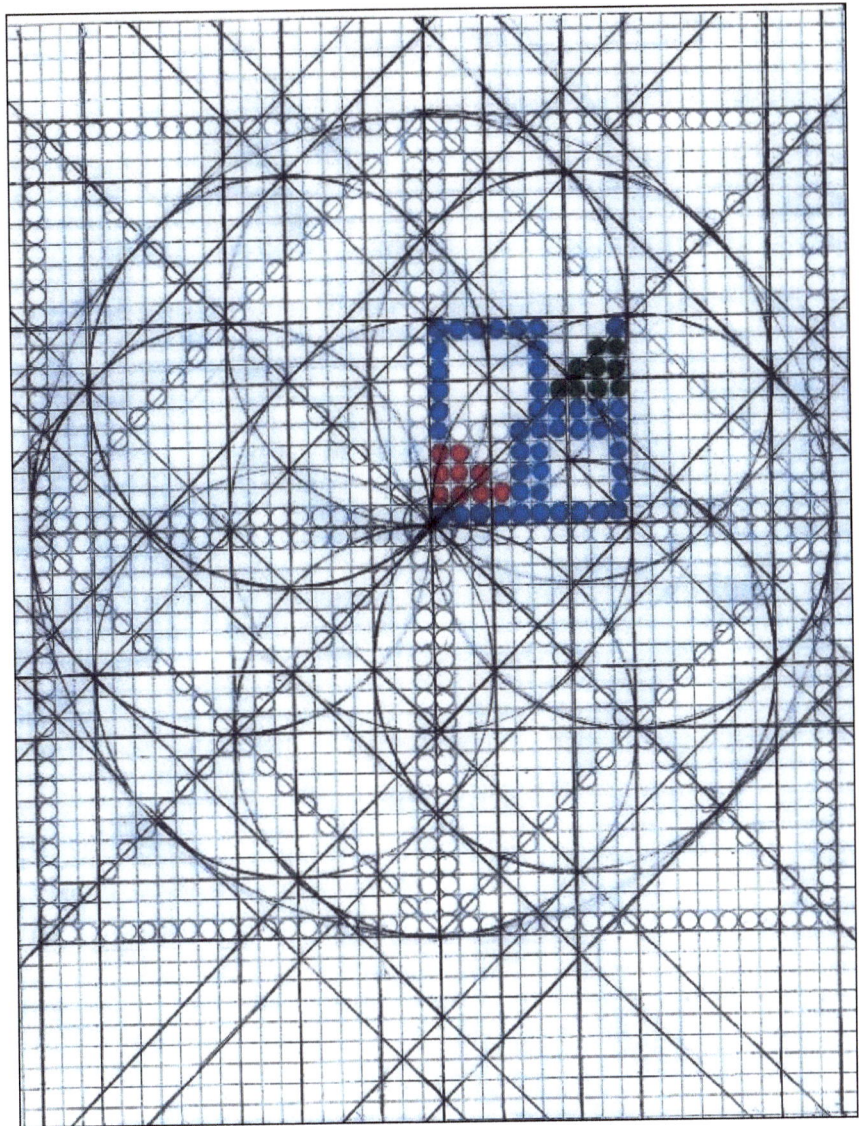

Figure 81

The green triangle now moves left one quadrant as shown in Figure 82.

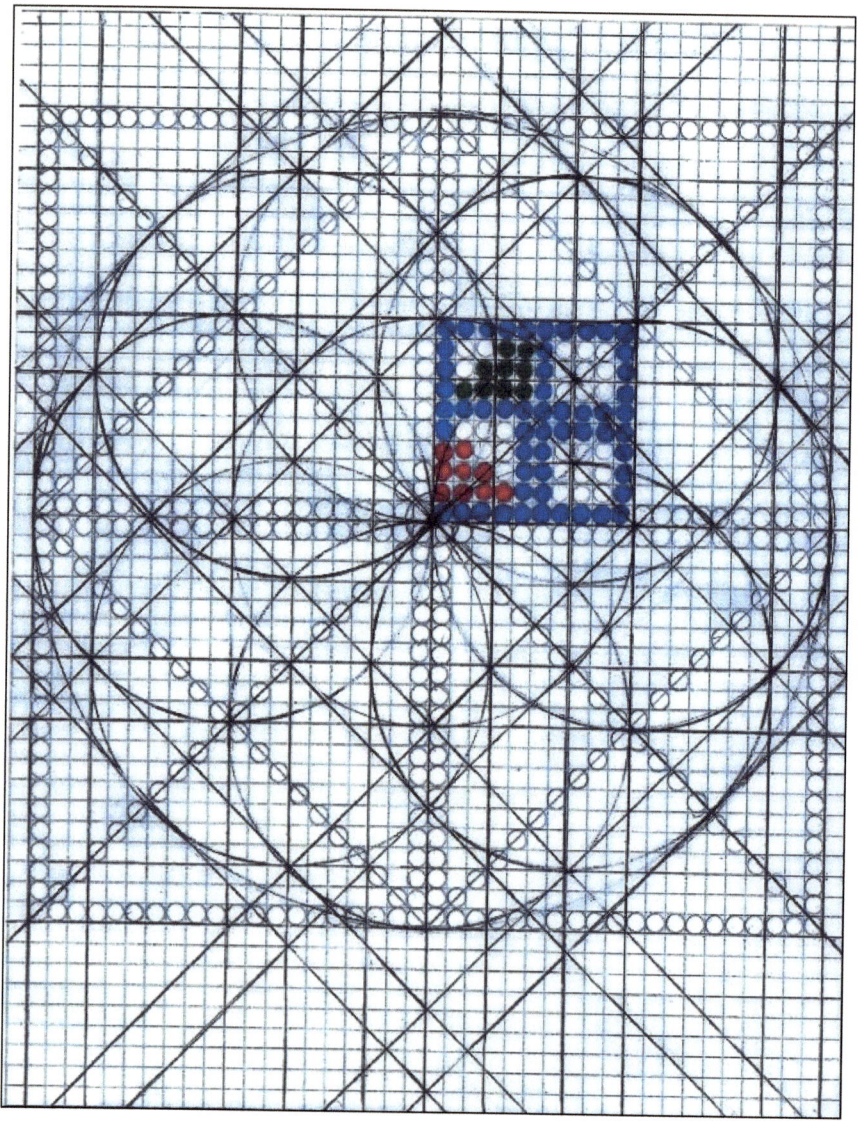

Figure 82

The red triangle now moves horizontally to its initial position as shown in Figure 83.

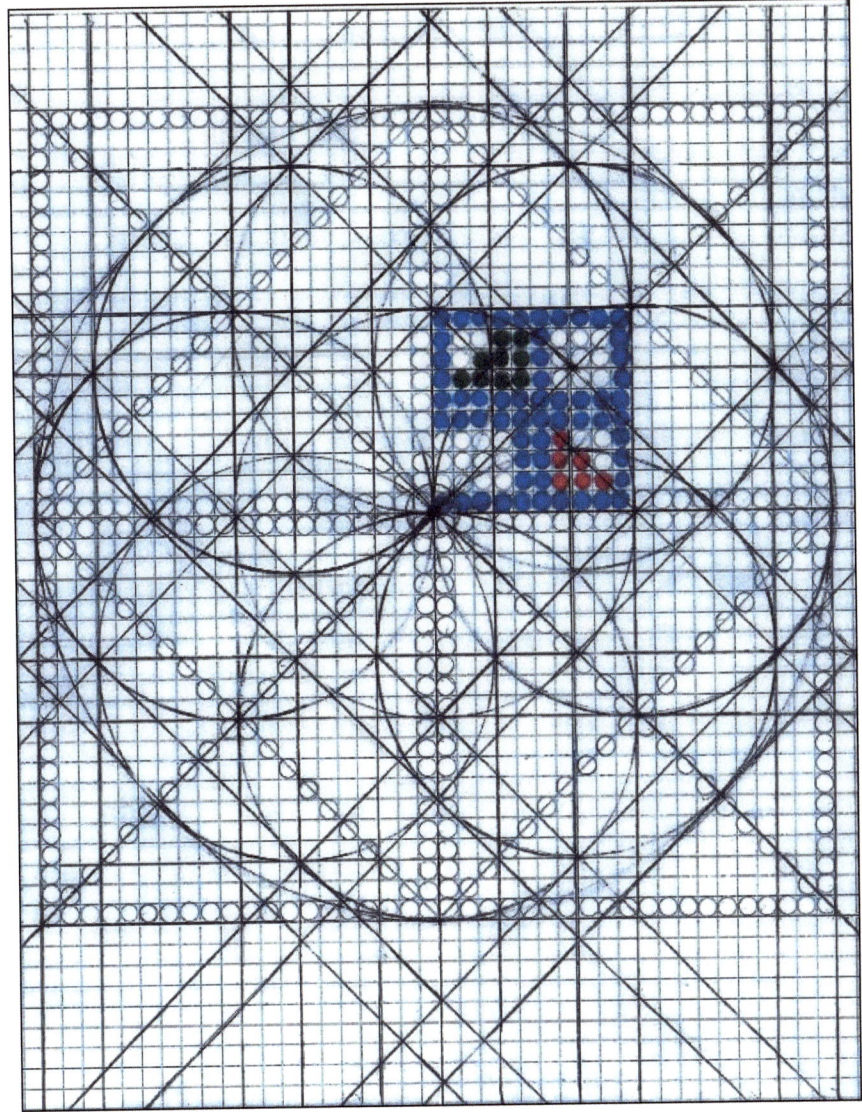

Figure 83

Figure 84 shows the final configuration after 4 moves of each triangle.

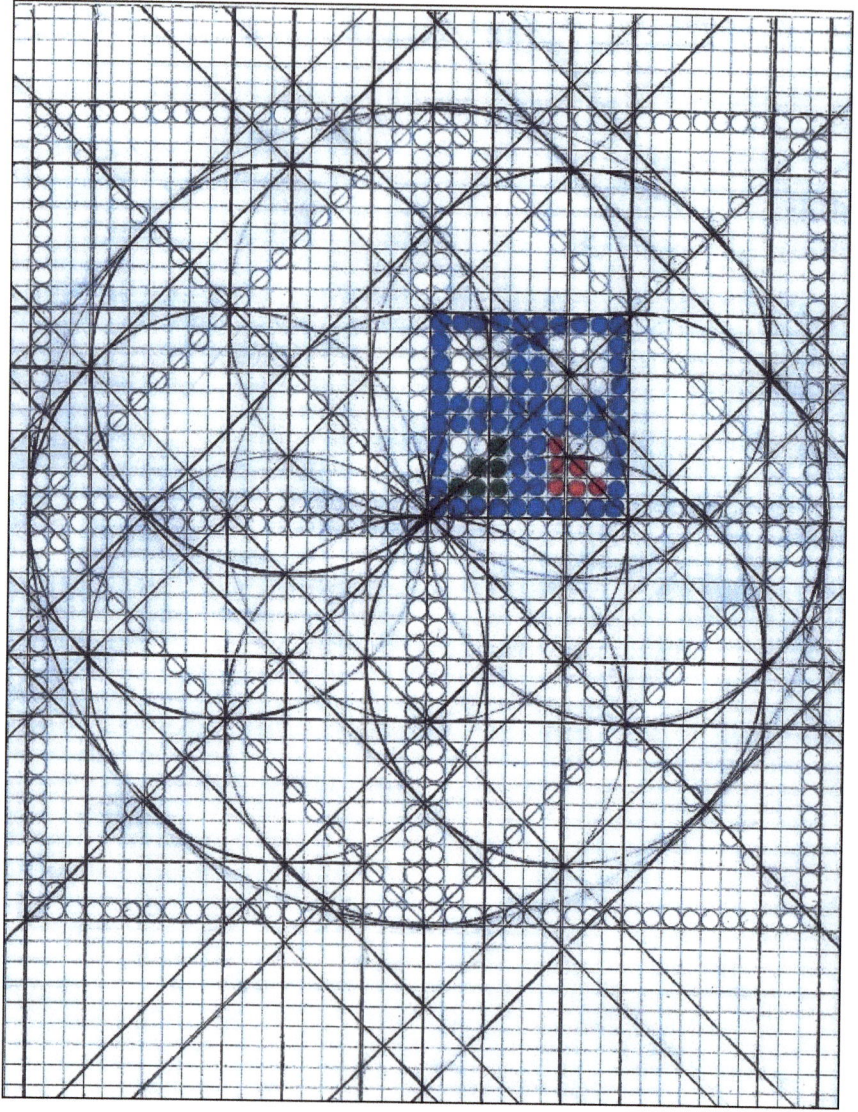

Figure 84

It is observed that when the red and green triangles are squared (A^2 and B^2) that $A^2 = 5 \times 5 = 25$, and $B^2 = 5 \times 5 = 25$, and $A^2 + B^2 = C^2$ or $25 + 25 = 50$ (Area). And $5 \times 4 = 20$ so $A + B = C$ or $20 + 20 = 40$ (Perimeter).

It is also observed that 3×3 squares are formed in the center of each quadrant. And $3 \times 3 = 9$ and $4 \times 9 = 36$ and $36 \times 100 = 360$.

The numbers are shown in Table 3.

Table 3:

$B_A + B_B = C_V (C_H)$
0 − 0 + 0 = 0 (0) $A_A + A_B = C_V (C_H)$
1 − 1 + 1 = 2 (0) 21 − 1 + 1 = 2 (0)
2 − 2 + 2 = 4 (0) 22 − 2 + 2 = 4 (0)
3 − 3 + 3 = 6 (0) 23 − 3 + 3 = 6 (0)
4 − 4 + 4 = 8 (0) 24 − 4 + 4 = 8 (0)
5 − 5 + 5 = 10 (0)   25 − 5 + 5 = 10 (0)

6 − 1 + 1 = 10 (2) 26 − 1 + 1 = 0 (2)
7 − 2 + 2 = 10 (4) 27 − 2 + 2 = 0 (4)
8 − 3 + 3 = 10 (6)   28 − 3 + 3 = 0 (6)
9 − 4 + 4 = 10 (8) 29 − 4 + 4 = 0 (8)
10 − 5 + 5 = 10 (10) 30 − 5 + 5 = 0 (10)

$A_A + A_B = C_V (C_H)$ $B_A + B_B = C_V (C_H)$
11 − 1 + 1 = 0 (2)   31 − 1 + 1 = 0 (2)
12 − 2 + 2 = 0 (4)   32 − 2 + 2 = 0 (4)
13 − 3 + 3 = 0 (6)   33 − 3 + 3 = 0 (6)
14 − 4 + 4 = 0 (8) 34 − 4 + 4 = 0 (8)
15 − 5 + 5 = 0 (10)   35 − 5 + 5 = 0 (10)

$B_A + B_B = C_V (C_H)$ $A_A + A_B = C_V (C_H)$
16 − 1 + 1 = 2 (0)   36 − 1 + 1 = 2 (0)
17 − 2 + 2 = 4 (0)   37 − 2 + 2 = 4 (0)
18 − 3 + 3 = 6 (0) 38 − 3 + 3 = 6 (0)
19 − 4 + 4 = 8 (0)   39 − 4 + 4 = 8 (0)
20 − 5 + 5 = 10 (0)   40 − 5 + 5 = 10 (0)

All numbers on the Perimeter are Zero based and the square has 4 quadrants and 3^2 (9) in the center of each quadrant.

Chapter 8
Pi (π), Protractor, Abacus

All numbers are a result of a Prime One configuration. There are infinite configurations. Pi (π) is a Base 3 configuration. All Zero-Based numbers are a circle within a square (N_Z). All Infinity-Based (N_I) numbers are a square within a circle.

π Base

The count begins at zero and continues to 5_Z horizontal and 5_Z vertical. The square is 3_Z x multiplier 3 and equals 9_Z. The whole number is 3_I its axis of rotation is the center of 1_I (∞). Note: The first Whole Number is 1_Z squared. 2_Z is not squared because the axis of rotation would be Zero. This is a Base 3 Numbering System. Reference Figure 85.

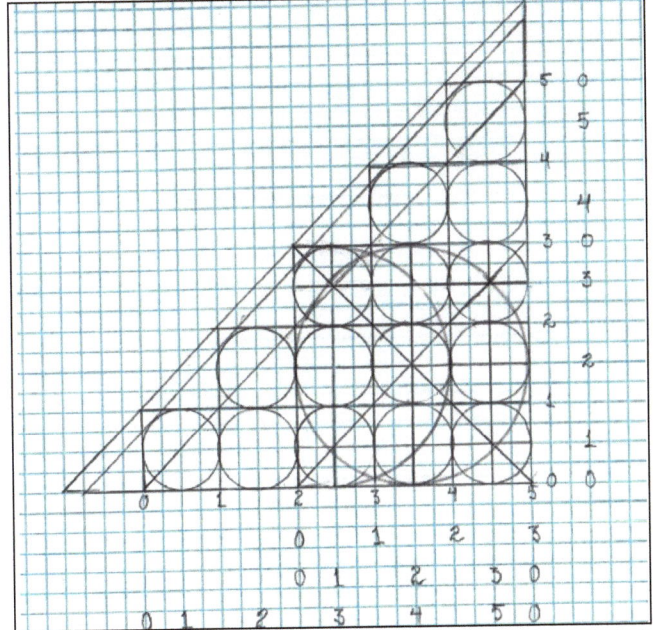

Figure 85

The Awakening of Numbers | 97

Figure 86 depicts 3 columns of 3_Z or 3 rows of 3_Z arranged in a square. The corners of the square do not contain numbers and the square does not rotate.

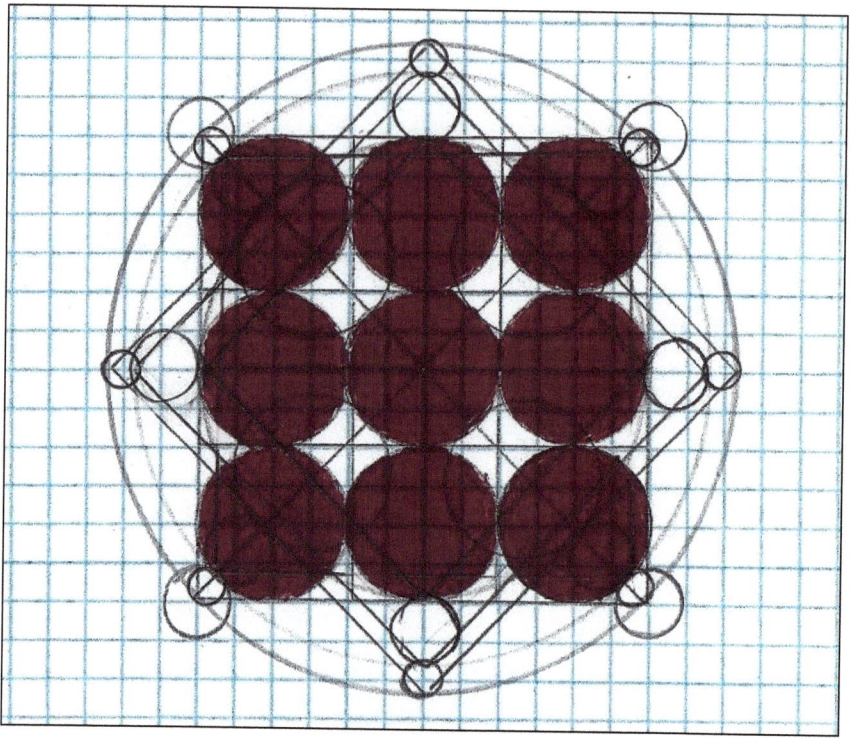

Figure 86

Figure 87 depicts the first expansion of the center 1_Z. The center Prime One is expanded to form the center Decimal 1_Z.

Figure 87

The Prime One continues to expand to the perimeter of the square forming 9_Z. This creates a vertical and horizontal diameter of 3_Z each and is depicted in Figure 88.

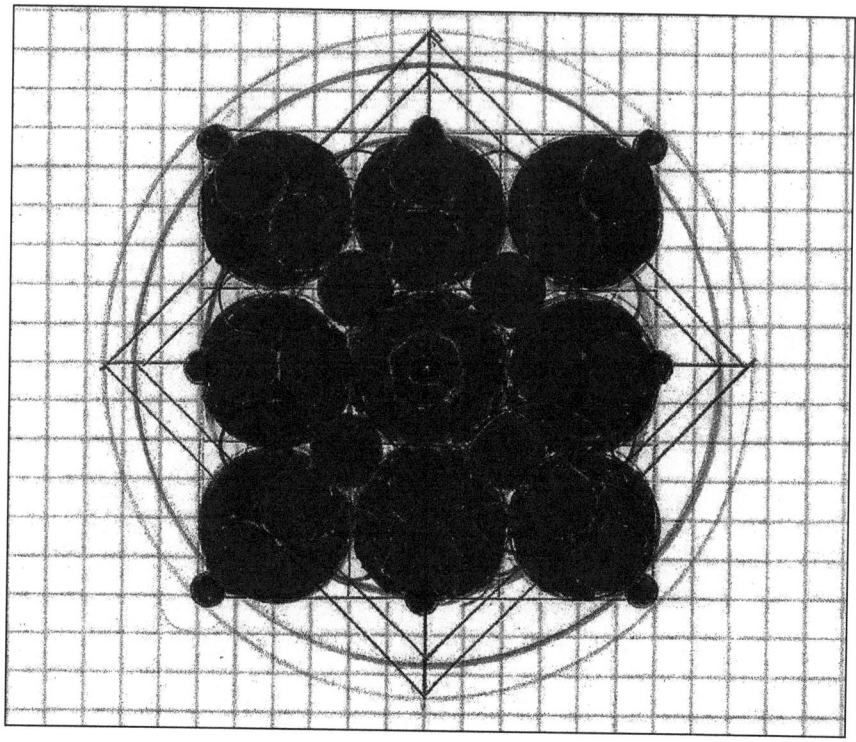

Figure 88

A circle within the 9_Z square is formed using the vertical and horizontal 3_Z Diameters, this circle does not rotate. The circumference of the circle enclosed within the 9_Z is completed with the insertion of a diagonal 1_Z within the enclosed circle for each of the 9_Z square quadrants. Each diagonal diameter or radius may rotate to a vertical or horizontal position but no further. This results in a 3_Z circle that is enclosed in the 9_Z square. Reference Figure 89.

Figure 89

Figure 90 shows a circle of 9_Z, this is $3_Z \times 3 = 9_Z$, or Diameter = 8_Z + the Axis $1_Z = 9_Z$. There are 4 diameters of 3_Z resulting in $3_Z \times 4 = 12_Z$, this is 3 squared and produces a square root of 3_Z, i.e.: $12_Z - 9_Z = 3_Z$. (Note: The Multiplication symbol used is •, indicating that the multiplication is circular vice linear (x). The 9_Z square perimeter count is $(1.5_{ZI} + 1.5_{ZI}) • 4 = 12_{ZI}$.

Figure 90 is 3_Z squared and is a product of a Zero-Based 3 number. The number is measured and counted with Base 10 (Decimal) symbols indicating it is contained with the Decimal System. The Decimal diagonal calculation uses a method like that used to determine a square root. The result of 3.141592654 is based on a calculation technique used at the time of development. 3.141592654 is 3_Z + $.141592654_I$. The measurement of the diagonal that results in 3.141592654 also results in designation of the square as π (Pi).

The count in a Zero-Based 3 system using Decimal System symbols is: 0, 1, 2, 3, 10, 11, 12, 13, 20, 21, 22, 23, 30, 31, 32, 33, 40, 41, 42, 43, 50, etc., the Pi count as currently presented indicates that Pi has been incorporated into the Decimal System as 1_π. Figure 90 depicts the incorporation of π within the Decimal Counting Board and the corresponding counts.

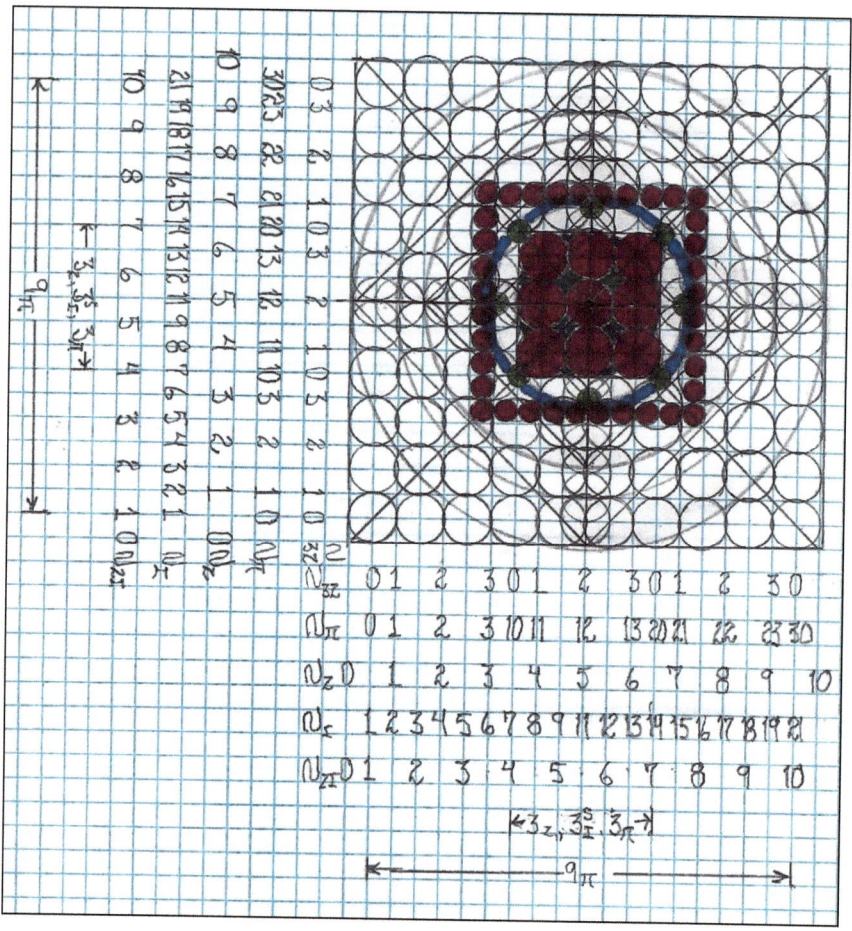

Figure 90

Figure 91 depicts a Circle with π contained within. The surrounding circle is divided into Hours. π does not rotate, the 3_{ZI} within the π square rotates.

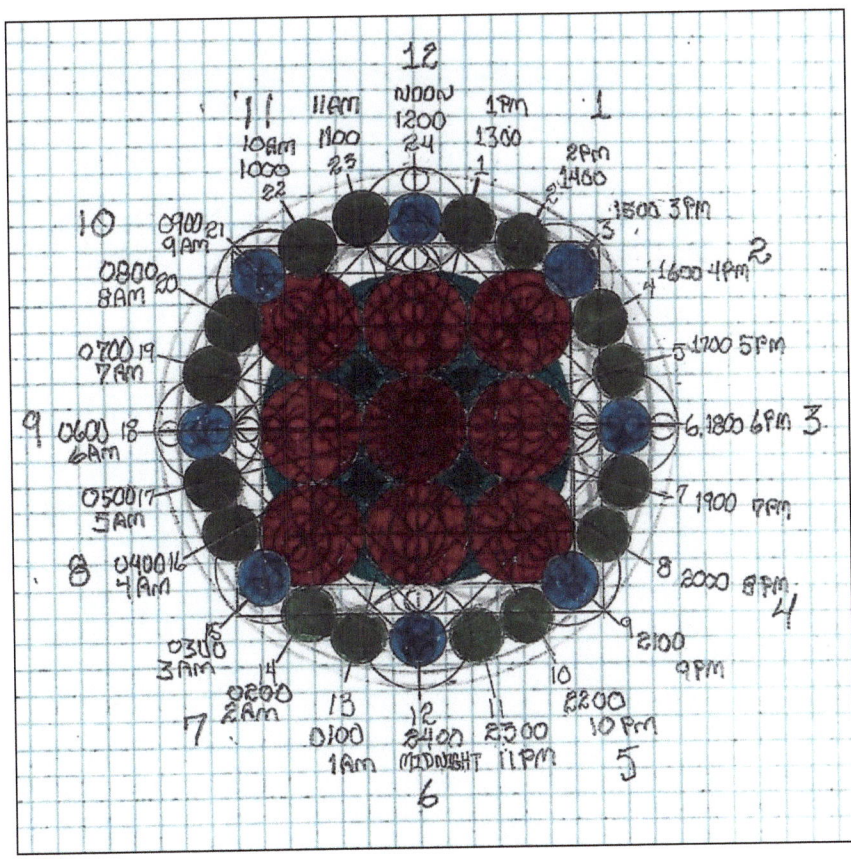

Figure 91

The diameters of the 3_{ZI}, enclosed within the 9_Z square, are expanded to 4_{ZI} ($3_Z + .5_I + .5_I$). The expansion results in the corners of π being enclosed within a $.5_I$. π does not rotate. Reference Figure 92.

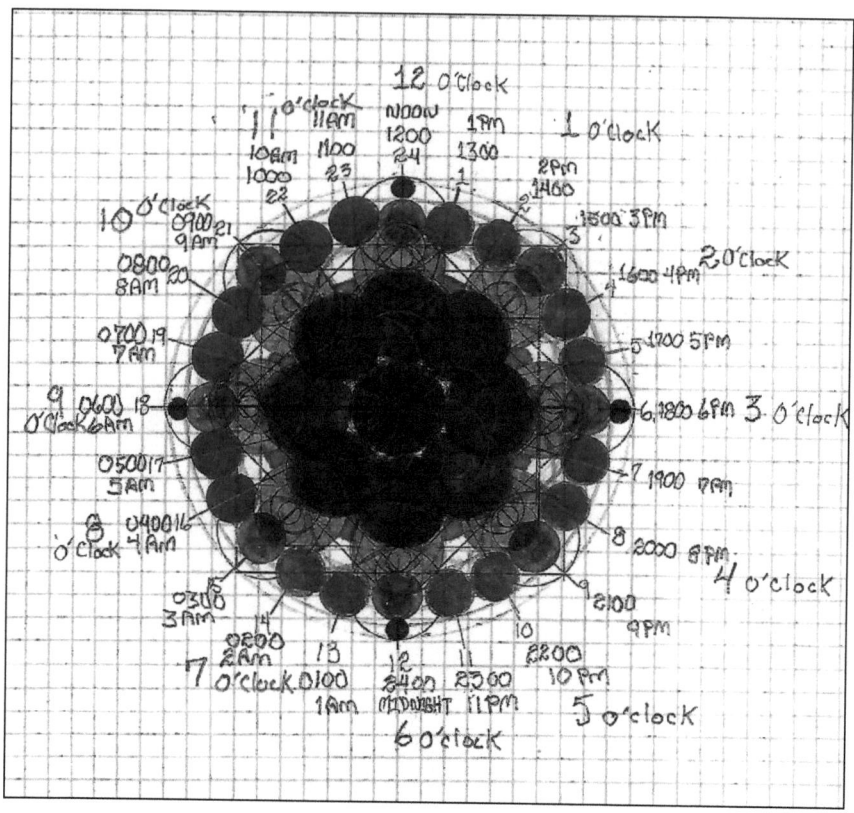

Figure 92

The 4_{ZI} diameters are converted to 3_I Radii resulting in 4 diameters of 6_I. π can rotate. π degrees are assigned (N_π). π degrees differ from Decimal degrees (N_{ZI}) in that the π degrees begin and end at Zero. $(0_\pi + 180_\pi) + (180_\pi + 0_\pi) = 360_\pi$. Reference Figure 93.

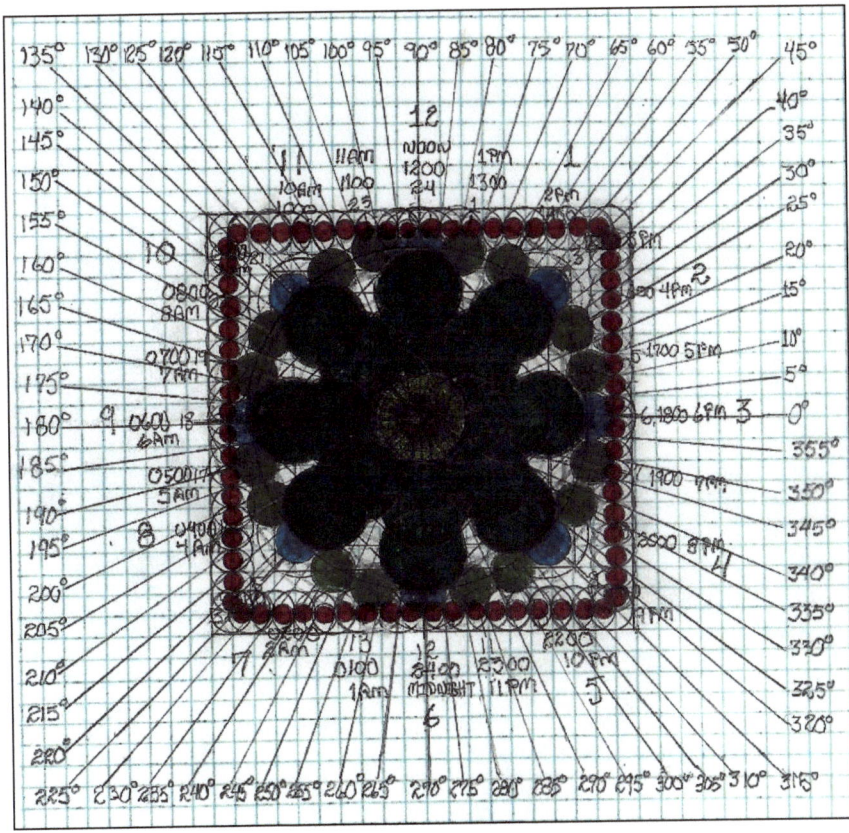

Figure 93

As π rotates each radius becomes the whole number 3_I resulting in rotatable diameters of 6_I each. ($4 \cdot 6_I = 24$ hours). Reference Figure 94.

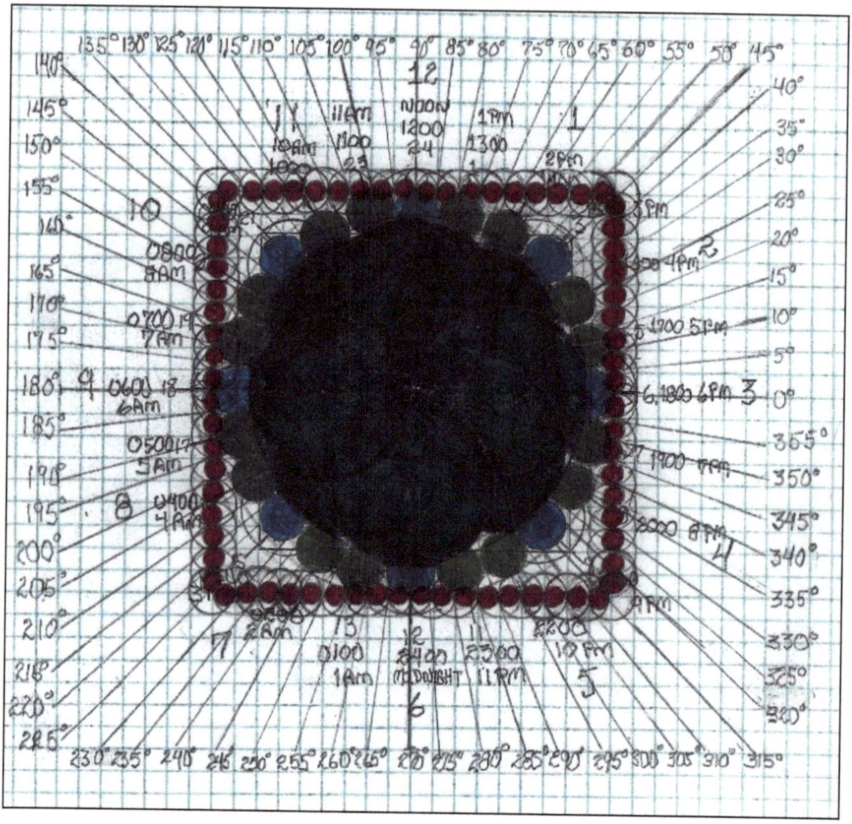

Figure 94

Protractor

Protractors are derived from a combination of π and the Decimal System ($5_Z + 5_Z$).

Figure 95 depicts the basic Zero armature for the construction of a Protractor.

Figure 95

Figure 96 depicts π which are contained in $5_Z \times 5_Z$ squares. Within each $5_Z \times 5_Z$ square is a $3_Z \times 3_Z$ square, within the $3_Z \times 3_Z$ square is 3_I and when 3_I is externally squared it becomes N_π. The count shown is $1_\pi, 1_\pi, 1_\pi, 1_\pi, 1_\pi, 1_\pi$.

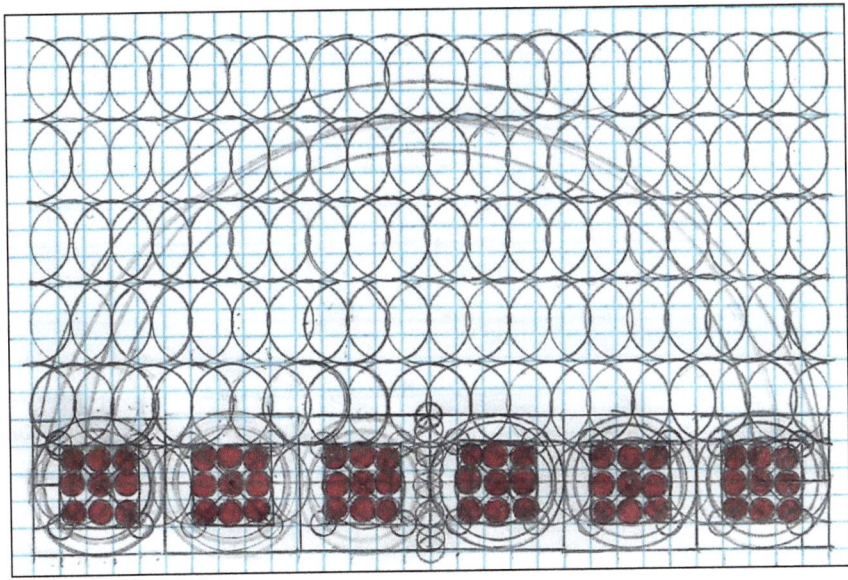

Figure 96

Figure 97 is the expansion of the N_π base showing the conversion of $N_{\pi Z}$ to $N_{\pi I}$ and it is enclosed within 5_I. A protractor armature is then constructed, consisting of 6 squares with 5_i, enclosed, horizontally and 3 squares of 5_I vertically. Each 5_I square is 1 square inch.

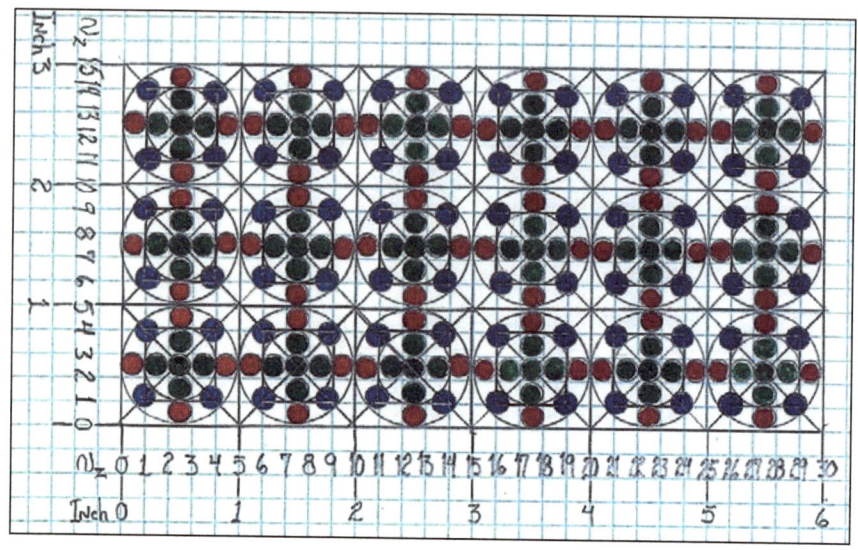

Figure 97

Decimal Protractor

Figure 98 shows a Decimal Protractor overlay of the protractor armature.

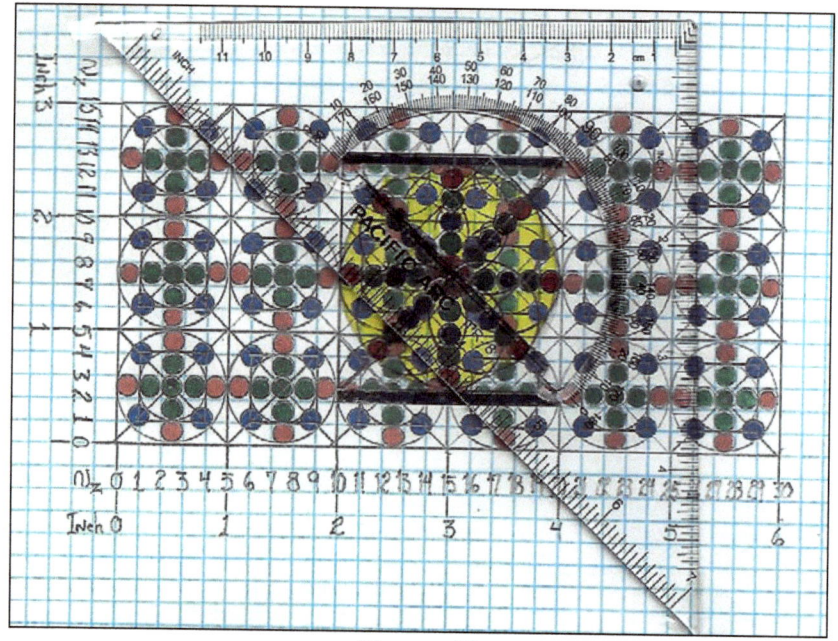

Figure 98

The Decimal Protractor rotates 180 , it is constructed as a half-circle. Figure 99 shows a squared circle with 2 diagonals with a count of 14.14213562_{ZI}.

When the diagonal is rotated to the horizontal the 14.14213562_{ZI} count is divided by 10 to produce the count of 1.414213562_{ZI} currently referred to as the Square Root of 2_Z. Figure 99 shows the rotation of the diagonal to the horizontal.

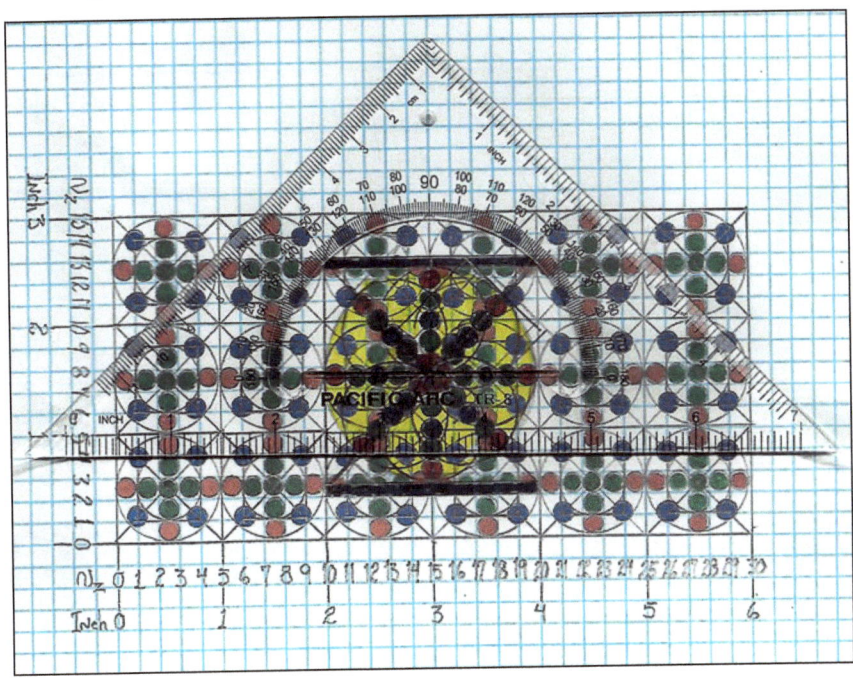

Figure 99

As the square is rotated a internal circle is created. The Decimal Protractor's 0-180 degree horizontal line can be placed at any point on the circumference of the inner circle, that was created as a result of rotation of the external square. This results in a relative measurement of the lines contained within the circle with a circumference that has no beginning or end. The rotation of the square also creates a circle external to the square. This Circle becomes Pi (π). Reference Figure 100.

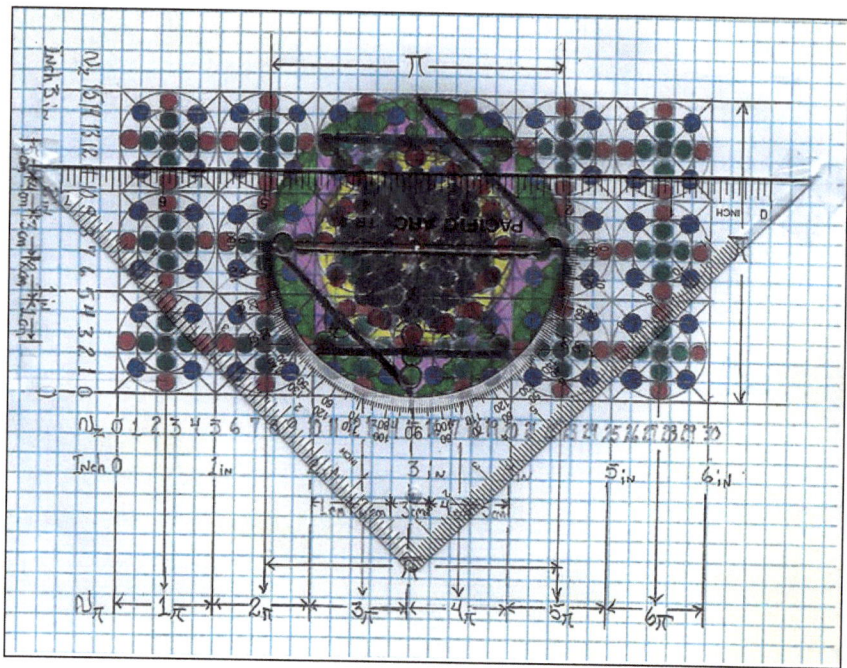

Fig. 100

The Protractor Armature is now configured to permit rotation of π. Each Base square is configured as $3_Z \times 3_Z$. The count center to center of each Base square is 6_π. The vertical and horizontal lines are configured as N_{ZI} resulting in a count of 10_{ZI}. This configuration will allow rotation of π. Reference Figure 101.

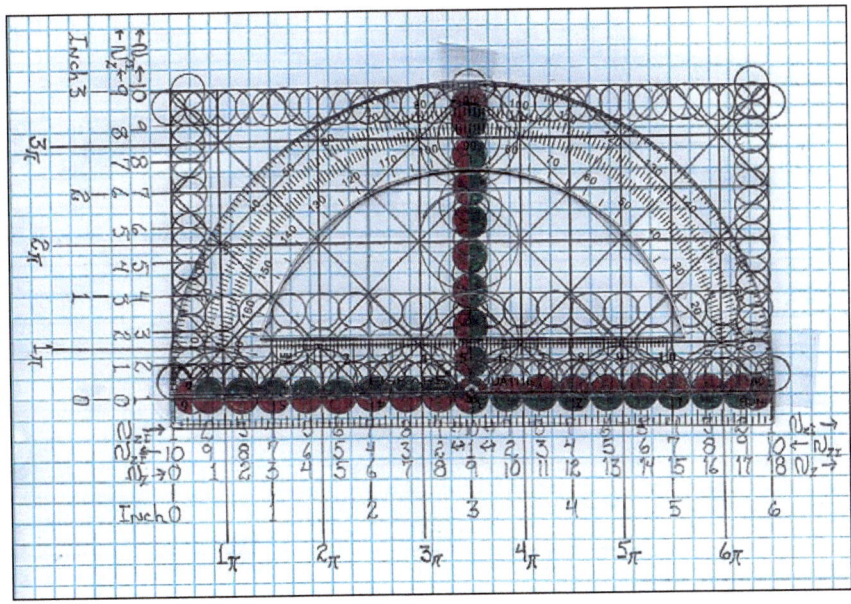

Figure 101

One half of the π Base on boths sides of of the center are rotated to the center vertical. Figure 102.

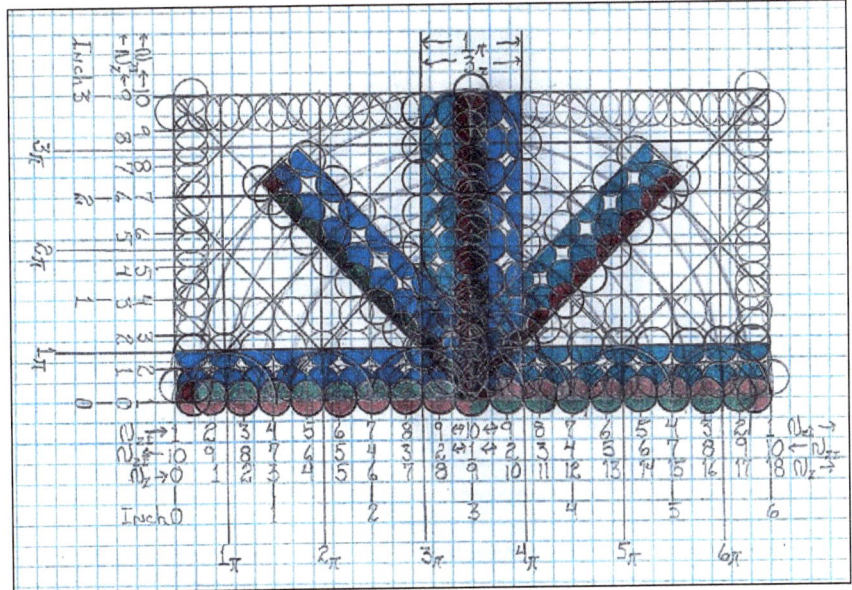

Figure 102

π is developed on the vertical with the 2 rotated horizontal half π bases. Reference Figure 103.

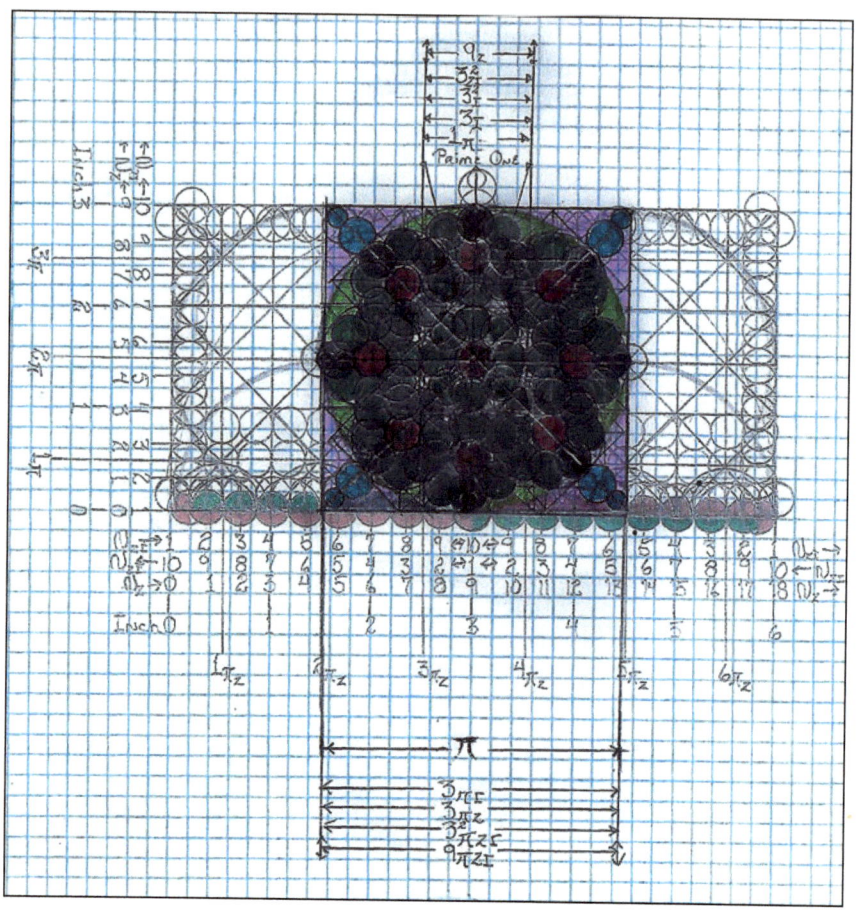

Figure 103

One half of Pi is rotated clockwise to the diagonal 45, the remaining half is rotated counterclockwise to 45 as shown in Figure 104.

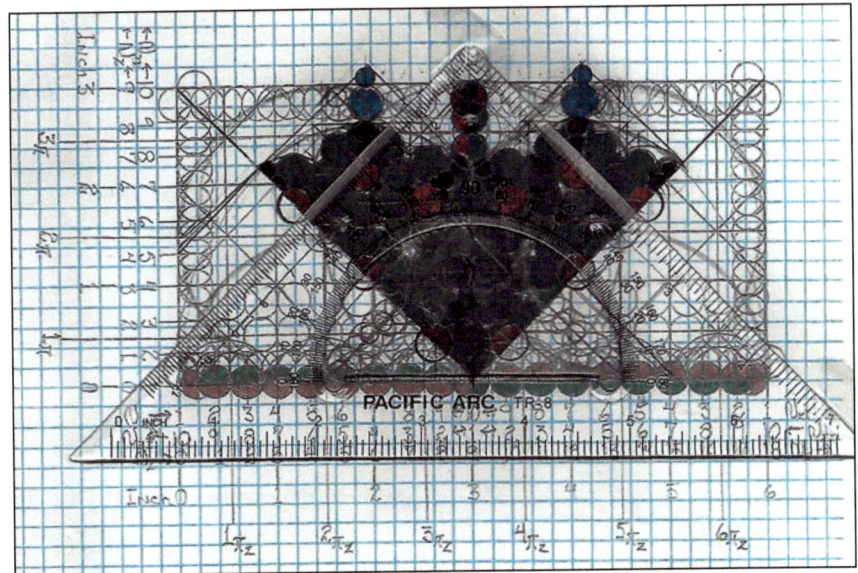

Figure 104

Each Pi half continues to the lower horizontal Base line. Reference Figure 105.

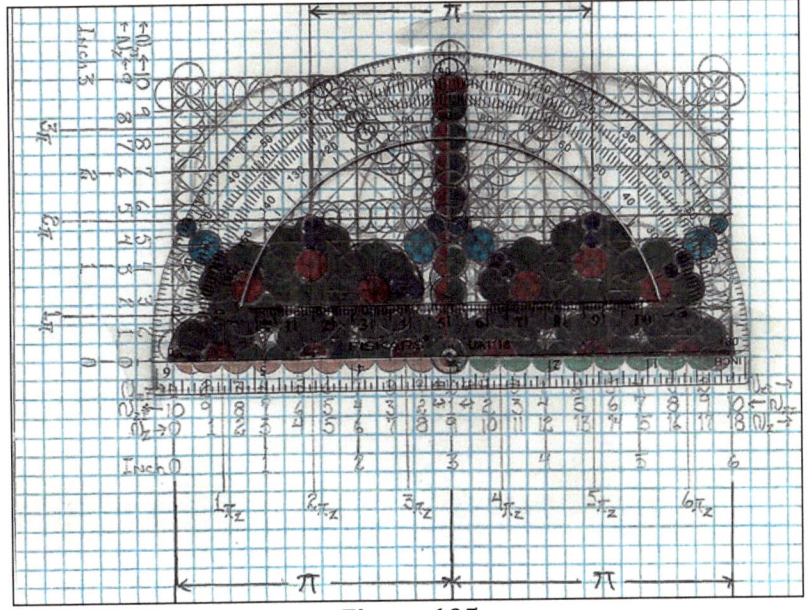

Figure 105

This is the Pi Protractor construct. It does not rotate as the Decimal Protractor construct does. The 0 – 180 may be placed either horizontally or vertically. The starting position of the line being measured may be either horizontal or vertical and rotates to the measured position. All lines are infinite in length and their rotational position is measured before application of a linear limiting value (segment). Any single line rotates 0 - 90 or 90 - 180 or 180 - 90 or 90 - 0. The entire construct and protractor may be inverted, and the measurement processes may then be performed and added to a previous measurement if desired. Understanding that the inverted construct and protractor 0 replacing 180 and 180 replacing 0. There is no 0 in a Decimal 360 circle.

The Chinese Abacus

Analysis of the Pi Protractor armature reveals that a Chinese Abacus construct is also present. The Chinese Abacus construct is most likely the origin and preceded the Pi Protractor armature. The rows of the Chinese Abacus armature are divided into 2_π circles with $½_\pi$ outside the armature on the vertical and horizontal edges. The left vertical ½ Pi is added to the right vertical $½_\pi$ creating 13 columns of 6_π. The bottom horizontal $½_\pi$ is added to the top horizonal $½_\pi$ creating 7 rows of $13_{\pi ZA}$ and 13 columns of $7_{\pi ZA}$. This is a Chinese Abacus. Reference Figure 106.

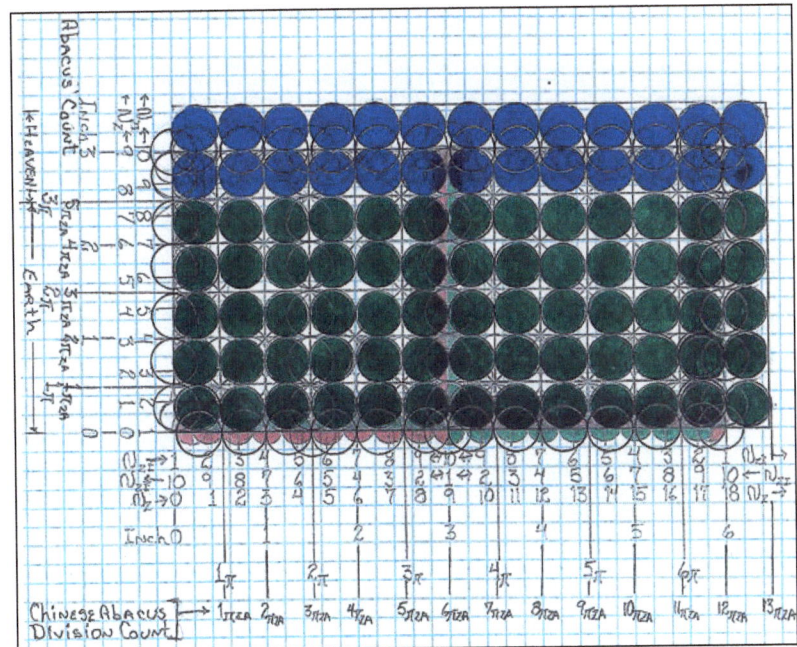

Figure 106

More Pi (π)

Figure 107 shows $\pi^2 + \pi^2$ within the Decimal System.

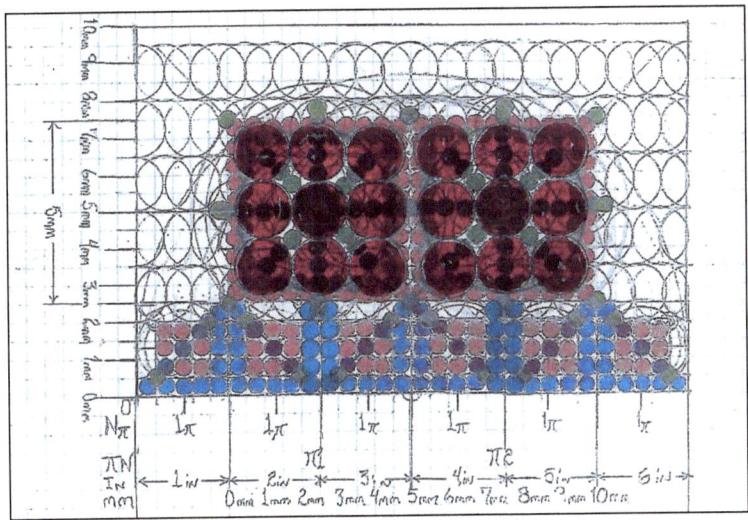

Figure 107

Figure 108 shows $\pi_I + \pi_I$ within the Decimal System.

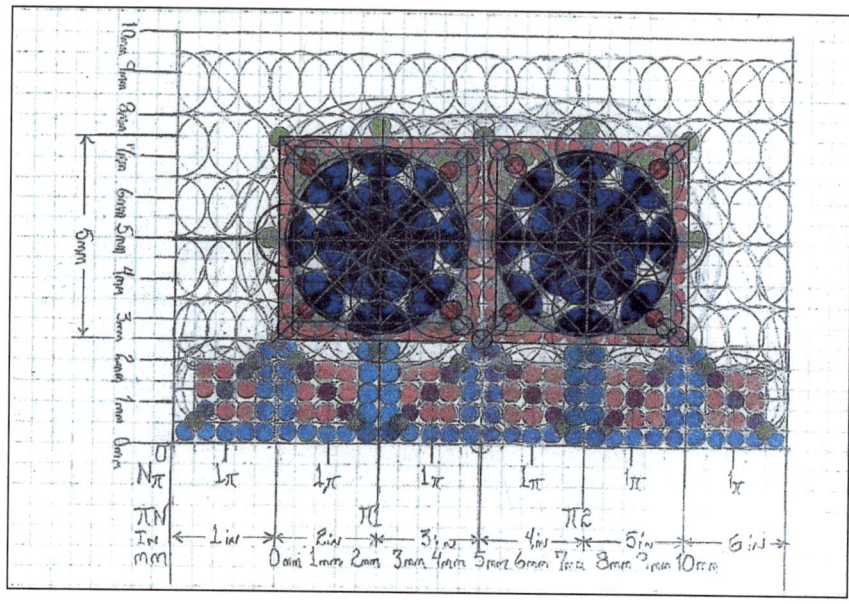

Figure 108

Figure 109 shows $\pi_{ZI} + \pi_{ZI}$ within the Decimal System.

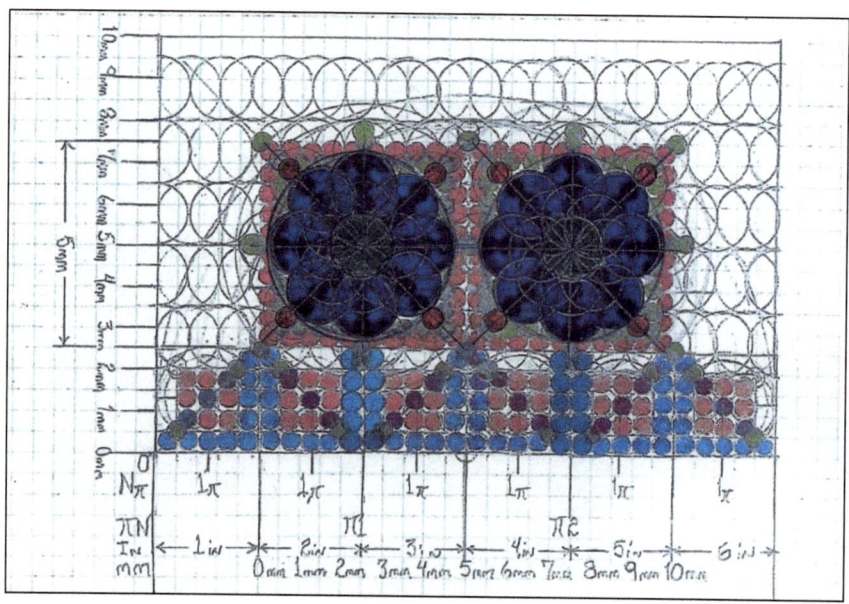

Figure 109

Figure 110 shows the Decimal System within $\pi_{ZI} + \pi_{ZI}$.

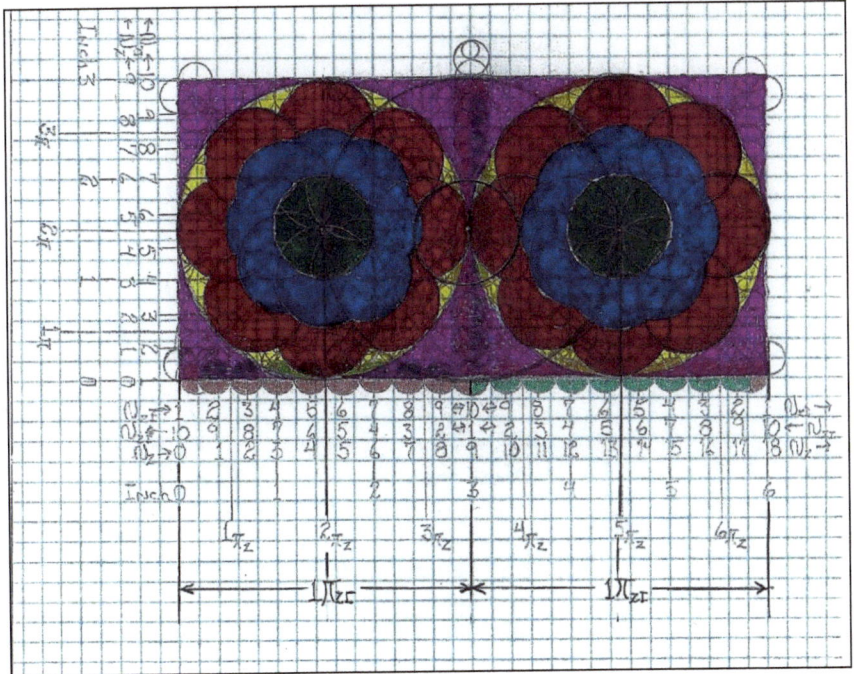

Figure 110

Chapter 9
Illustrations

1. Linear Number

2. Circular Number

3. Infinity-Based Number

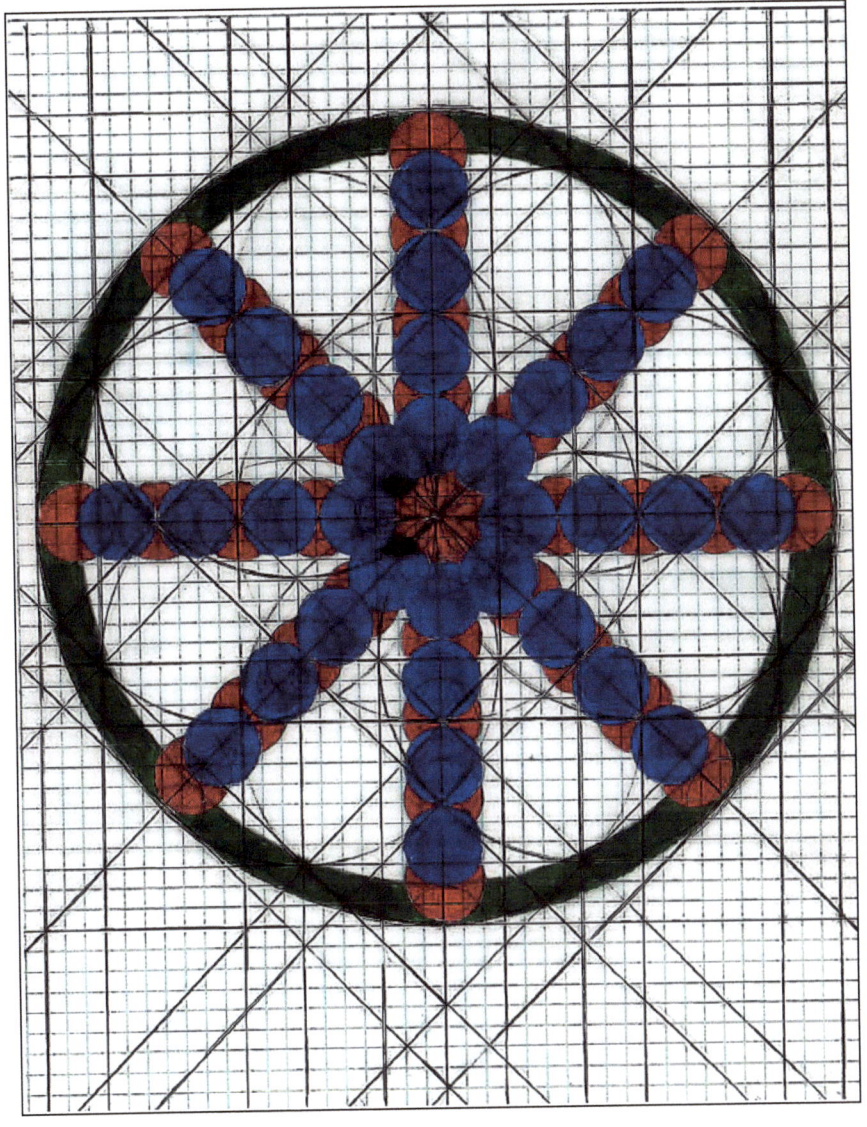

The Awakening of Numbers | 121

4. Squared Number

5. Squared Numbers

6. Squared Infinity-Based Number

7. Number Rose

8. Whole Number

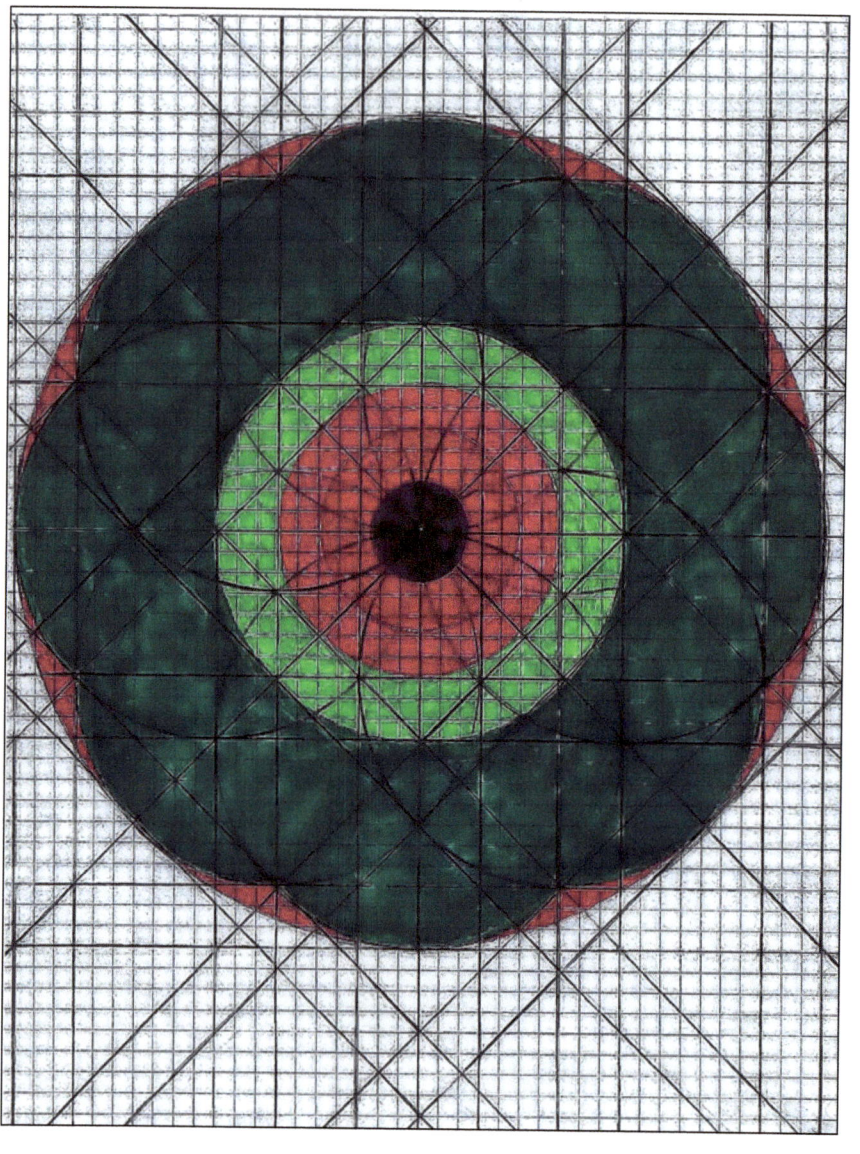

9. Decimal Zero-Based Numbers

10. Decimal System Infinity-Based Number

11. Diagonal Decimal Number Squared

12. Measuring Squares

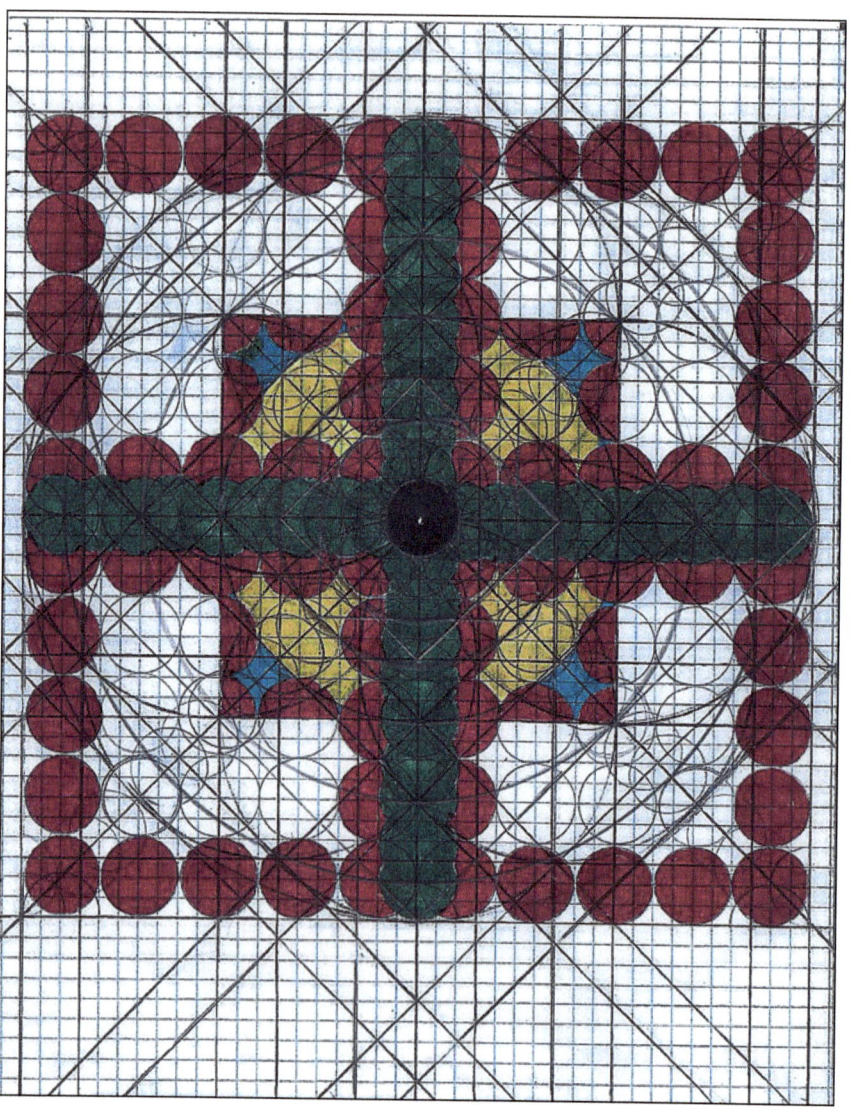

www.ingramcontent.com/pod-product-compliance
Lightning Source LLC
Chambersburg PA
CBHW041940240526
45473CB00033B/26